Synthesis Lectures on Engineering, Science, and Technology

The focus of this series is general topics, and applications about, and for, engineers and scientists on a wide array of applications, methods and advances. Most titles cover subjects such as professional development, education, and study skills, as well as basic introductory undergraduate material and other topics appropriate for a broader and less technical audience.

David K. Ferry

Quantum Information
in the Nanoelectronic World

 Springer

David K. Ferry
Arizona State University
Tempe, AZ, USA

ISSN 2690-0300   ISSN 2690-0327 (electronic)
Synthesis Lectures on Engineering, Science, and Technology
ISBN 978-3-031-62924-2   ISBN 978-3-031-62925-9 (eBook)
https://doi.org/10.1007/978-3-031-62925-9

This Springer imprint is published by the registered company Springer Nature Switzerland AG
The registered company address is: Gewerbestrasse 11, 6330 Cham, Switzerland

If disposing of this product, please recycle the paper.

If we were not able or did not desire to look in any new direction, if we did not have a doubt or recognize ignorance, we would not get any new ideas....scientific knowledge today is a body of statements of varying degrees of certainty. Some of them are most unsure; some of them are nearly sure; but none is absolutely certain. Scientists are used to this.

—*Richard P. Feynman*[1]

[1] Feynman, R. P.: The Meaning of it All: Thoughts of a Citizen-Scientist. Reading, MA, Addison-Wesley, 1998.

Preface

This book has arisen from a chapter written for a long volume on semiconductor devices.[2] When Springer asked me to extend this chapter into the present book, I wasn't sure this was something I wanted to undertake. A few days later, however, I was driving across west Texas and being quizzed for some hours by one of my daughters about quantum information. She recently had become an Associate Vice President of Research at Arizona State University, and felt her background in biology needed some expansion for her new job. After that, it finally occurred to me that there might be many other people in the world in a similar situation to her, and this could provide a considerable market for such a book.

Quantum mechanics is a relatively young science, having appeared less than a century ago. As a result, there remains considerable debate, and certainly confusion, about what it all means. Even Feynman was wont to say that "...nobody understands quantum mechanics." Yet, here we are, trying to initiate a quantum information age, with governments around the world spending enormous amounts to push the field forward. One might reasonably note that if the scientists can't agree on what quantum mechanics means, or how it should be applied, how can the interested bystander understand what it all means? So, it might be good to go through how and why this confusion exists. First, though, there is a caveat: I am an engineer and my toolbox tends to contain theories and experiments that explain what works for engineers. I have a good friend, who is a professor of physics in Barcelona and who really likes to get into the philosophical underworld (or ontology as it prone to be called) of trying to understand quantum mechanics. Yet, in the end, he (like me) tends to follow what works for engineers. We have published papers and books together, and it is this view of the world that will color most of the discussion in this book.

---

[2] Ferry, D. K.: "Nanoelectronic Systems for Quantum Computing." Handbook of Semiconductor Devices, ed. by Rudan, M., Brunetti, R., Reggiani, S., Springer Nature Switzerland, 2023. Chapter 33.

The quantum world began on December 14, 1900, when Max Planck gave a talk at the monthly meeting of the German Physical Society (as it is called today). He introduced quantization of the photon in order to describe a new theory of electromagnetic radiation from hot bodies. This didn't come out of the blue, as there were new experiments being carried out in Berlin that showed that the old theories were wrong. To reach the "right" theory, he had to introduce his relation $E = hf = \hbar\omega$, where the first term is the energy of the photon, and the last two terms involve moving a factor of 2π from h to f. This, by itself, doesn't exhibit the quantization. It is when one considers the light-wave power reaching a surface that one sees that the light arrives as so many photons per second— light must be a particle. This view wasn't new, as the ancient Greeks thought that light was particles and even Newton held this view. But, experiments in the late 1700s showed interference of light waves and this led to the view that light was an electromagnetic wave (and not a particle). Of course, just a few years later, Einstein showed that experiments on the emission of electrons from solids when illuminated by light (of sufficiently high photon energy) were explained by Planck's new quantum theory. Which was it to be; particle or wave? It certainly did not clear up the situation when Taylor, in 1909, showed the answer was "both." He studied very weak light in which only a single photon at a time could be passing through the interference region, and found that the photon appeared on the photographic plate as a particle, but when an ensemble of photons had passed, the interference pattern would emerge. Now the conundrum was established; light was paranoid. Some years later, electrons were shown to have the same paranoia.

Attention then turned to the world of atomic structure. Bohr, following ideas from his mentor Rutherford and from the Japanese scientist Nagaoka, developed an atomic model in which Planck's quantization led to quantization of the angular momentum of electrons circling the nucleus. In a subsequent paper, he did the reverse and showed that quantization of the angular momentum led to quantization of the electron energy, thus completing the tautology. But, the real quantum theory arrived in the mid-1920s with work by Heisenberg, de Broglie, and Schrödinger. Notably, it was de Broglie who asserted that each particle was accompanied by a "pilot" wave, and this combination would exhibit both wave and particle behavior. The schism arose from Bohr introducing his positivism (nothing exists until it is observed) on top of Heisenberg's matrix theory. This splits the science world into two parts: one camp involving mainly Bohr and Heisenberg (and their large number of acolytes), and a second involving Einstein, Schrödinger, and de Broglie. While these became the main antagonists in the debate that extends even until today, many other scientists fit themselves into either camp (although mostly the first camp, known as the orthodox or Copenhagen camp, for a variety of reasons). Qualities such as determinism and causality disappeared from Bohr's world view. The consequences of the debate and the influence the leaders had on the interpretations of quantum mechanics appear at several points in this book, but do not seriously hinder the advances which have been made in the quantum information field. This is perhaps because these advances depend upon results, and not particularly upon the question of how they occurred. Anyone who wants a deeper discussion of the history and debates may turn to my book *The Copenhagen Conspiracy*.

Quantum physics was involved in the nanoelectronics world almost from the first transistor. Its importance was discussed in a series of papers John Barker (a colleague from Glasgow) and I wrote that were published in the late 1970s on small, and very small, devices. Then, a significant review of the electronic properties of two-dimensional systems at low temperature by Ando, Fowler, and Stern appeared in 1982. So, the importance of quantum mechanics in semiconductor devices was clearly established, and many studies at low temperature were to follow. Yet, the industry ignored most of this work, pushing forward by empirical methods. These were based upon a scaling theory in which all dimensions were uniformly reduced to maintain the same electrostatics, and this worked well. Until it didn't! The problem with just reducing the size is that edge effects soon become a very large part of the problem. By the late 1990s, quantum physics was a major concern within the industry as new dielectrics, new structures, new materials, and the introduction of strain all required more knowledge about the science of quantum materials and the need for quantum mechanics in the understanding of how modern devices should work.

At about the same time, in the late 1970s and early 1980s, the first ideas for quantum computing began to appear. As is the case in most new scientific endeavors, some of the early ideas were either naive or somewhat misguided in their interpretation of quantum physics. And, this led to more debates in this new field. But, technology pushed forward. D-Wave launched its first quantum computer in 2011, and IBM put their first 5 qubit machine online around 2016 for scientists and industry to begin evaluating the new technology. However, qubits for quantum information is a somewhat older concept, as the suggestions for using quantum mechanics in optical communications dates to 1960. But, the real growth came later with the development of actual qubits. And, as one says, the rest is history, as quantum information has mushroomed both in algorithms and in technology since these beginnings. It is the current state of this new quantum information world that is the subject of this book, and hopefully this will be useful to the community.

It remains clear that a rosy future for quantum information is not accepted by all, even among the scientists. Most of these questions deal with quantum computing. While Chaps. 2–7 deal with the current state of the technology, the questions being asked are addressed in Chap. 8, where an attempt is made to look forward to just what the future may hold. Science progresses through propositions that are then examined by other scientists as well as the proposer. There are many examples in science where the proposition was a step too far, and the naysayers proved to be more correct. But, this approach is the heart of the scientific method. Even in the quantum world, we should not seek to eliminate the importance of the method.

I would like to thank Charles Glaser, my editor at Springer, for pushing this project my way and for getting it initiated. Thanks also go to Lara Ferry, Jon Bird, Gil Speyer, and David Wahl, all colleagues who volunteered to read the various drafts and provide exceedingly useful feedback to strengthen the book. Finally, thanks to my wife who has allowed me to invade her space by working at home these past few years.

Tempe, AZ, USA   David K. Ferry
October 2023

Contents

About the Author

David K. Ferry is Regents' Professor Emeritus in the School of Electrical, Computer, and Energy Engineering at Arizona State University. He moved to ASU in 1983, following shorter stints at Texas Tech University, the Office of Naval Research, and Colorado State University. He received his doctorate from the University of Texas, Austin, and spent a postdoctoral period at the University of Vienna, Austria. His research is focused on semiconductors, particularly as they apply to nanotechnology and integrated circuits, as well as quantum effects in devices and materials. In 1999, he received the Cledo Brunetti Award from the Institute of Electrical and Electronics Engineers, and is a Fellow of this group as well as the American Physical Society and the Institute of Physics (UK). He is the author of some 40 books and book chapters and more than 900 refereed contributions.

Introduction

The necessity for the creation of a calculating "machine" by which to assist with solving complicated calculations is not new, nor is the idea of assisting this process through quantum mechanics. Calculating machines date back centuries, with the perhaps earliest being the counting board and the slightly more modern abacus, which seems to have their roots in the ancient Greek and Roman times. These, however, were merely for counting, when traders needed to keep track of values that exceeded what could be quantified using only fingers and toes. More complicated calculations still needed to be done in vivo, or in our head. Concepts of modern "computing machines," in which calculations were done for the user, date at least to Charles Babbage in the mid-nineteenth century [1]. The understanding of *how* to compute came later with a construction provided in 1934 by Alan Turing [2]. Even with the application of quantum mechanical processes to computing, which appeared only a few decades later [3, 4], computer development has relied upon the Turing machine. And, while much has changed in the modern computing world, in terms of the size, speed, and capabilities of computing machines, they are still based upon the principles of the Turing machine and the concept of computable numbers (i.e., those numbers are said to be computable with finite resources, discussed more later).

Computational "power," the aforementioned size, speed, and capabilities of a computer, is limited almost exclusively by the technology that provides the internal circuitry of the computer. Just as the adaption of water power to machines and the arrival of the steam engine gave rise to the Industrial Age, the arrival of the integrated circuit in the late 1950s [5] led to the creation of what has been called the Information Age [6]. Today's computers are composed of a variety of microchips, that are integrated circuits composed of diodes, transistors, and even central processing units of the computer, all on a single silicon chip, and billions of these chips are produced every day by the technology companies. Indeed,

© The Author(s), under exclusive license to Springer Nature Switzerland AG 2025
D. K. Ferry, *Quantum Information in the Nanoelectronic World*,
Synthesis Lectures on Engineering, Science, and Technology,
https://doi.org/10.1007/978-3-031-62925-9_1

the integrated circuit affects just about every detail of our life today. This approach to integrated circuitry has followed two complementary paths for advancement, one focused upon adding more and more components to a single chip, and a second focused on making these individual components smaller and smaller through miniaturization. The subsequent advancements in microchip technology, and the resulting computer power, has been characterized over the years by Moore's Law [7], which posits that the number of transistors on an individual microchip will double every two years. When added to the possibilities given by the additional capabilities that are offered by quantum mechanics, there has been a growth in the area that today is called quantum information. This covers not only computing, but also communications, and especially secure communications, as well as sensing.

In this, and the subsequent chapters, the exploration of the realm of these various components of quantum information as well as the impact of nanotechnology in the area, will be the goal. But, it is important to note a caveat: Each of the areas mentioned above involve the concept of a measurement, in which some property must be quantified. Even the concept of a quantum mechanical process must involve some form of quantification of the results. This is most evident in quantum sensing, where the signal is being "sensed" in a manner that may be regarded as a measurement. However, it is unfortunate that today there is no single, universally accepted, ontological description of quantum mechanics [8]. That is, there is a variety of differing interpretations based upon differing opinions of the meaning of the measurement process, and in what constitutes a measurement. It is not the purpose here to engage in these arguments. Rather, the view of a measurement used here will that of a (hopefully) rational engineer trying to move forward with real quantum systems [9].

In this chapter, as the name suggests, we proceed with an introduction to the current state of modern nanotechnology, specifically nanoelectronics, and the world of supercomputers is presented. Along the way, the ideas of what makes a Turing machine, in terms of technology, are discussed. We then turn to the basics of quantum computing and its state today. This is followed by similar discussions on quantum communications and quantum sensing. The level of these discussions is only an overview, mainly to express the *what* and the *why* of quantum information today. This will include a discussion of some network and circuit principles important to the three parts of quantum technology. Then, in Chap. 2, the *how* will be investigated. In other words, in this second chapter, a deeper investigation will be made into key scientific elements that are more or less consistent across the entire landscape of quantum information. Finally, in the remaining chapters, more scientific details of the various technologies found in quantum information systems will be discussed.

1.1   Modern Nanoelectronics

As mentioned, the integrated circuit and Moore's law arrived in the late 1950s, some seven decades ago. Since then, the evolving computational ability of the integrated circuit has followed a generational development cycle in growth of the number of transistors, or gates, on a single integrated circuit, or *chip*, doubled gate count each generation. This doubling, or temporal "length," of a generation is only 18–24 months. As a result, even the chip in a modern cell phone has orders of magnitude more computing power than the supercomputers of just a few decades ago. At the current time, the newest leading-edge chips are known as the 3 or 5 nm generation.

While such a dimensional nomenclature no longer applies to the most critical dimension, which is the channel length in an individual transistor, it is still used to indicate progress being made as the generations move forward. That is, the channel length remains much larger than 3–5 nm, and the terminology is more a label, than a description. While it seems improbable that progress would continue at the rate prescribed by Moore's Law (after all, the distance between 2 Si atoms is ~0.25 nm), we still move on with this scalability of the technology. And, to a large extent, this is achieved through the down-scaling of critical lengths in the transistor [10]. Such a scaling process varies lengths in order to maintain the electrostatic topology of the transistor. However, there have been a number of paradigm shifts, which have taken us from the planar Metal–Oxide–Semiconductor Field-Effect Transistor (MOSFET) to the tri-gate or FinFET, to nanowire FETs, and to nanosheet FETs. In Fig. 1.1, both the FinFET and the nanosheet FET are depicted as an illustration of modern transistor design [11]. In this figure, the green parts are the silicon (Si) of the active channel(s), with conduction near the interface between this Si and the silicon dioxide (SiO_2) surrounding it.

Today, a chip may contain more than 10 billion transistors that contribute to many processing *nodes* (a computing unit on the chip). Often, each node may be capable of processing two or more *threads* simultaneously, so that the computing power of a chip is related to its parallelism, the product of the number of nodes and the number of threads per node. The chip may even have a few neurally-inspired processors that may be used for image processing or even AI learning applications. Accessing a computer-based (or controlled) system, such as a cell phone or a secure building, with facial recognition or finger prints are just two examples of pattern recognition, accomplished with these neurally-inspired processors. Or consider a modern car that may have dozens of chips in the various control circuits (and radios) that perform crucial tasks for the control processes that make the car function in a safe, and predictable, manner. Even the routine task of shifting gears often is now handled by a control computer and servos for the actuation process.

Moore's Law. Moore's Law is often cited as a technology law, simply because the most evident usage is the scaling rules that govern the downsizing of the transistors [10]. There are other technologies that follow such evolutionary laws, so this is not unique to

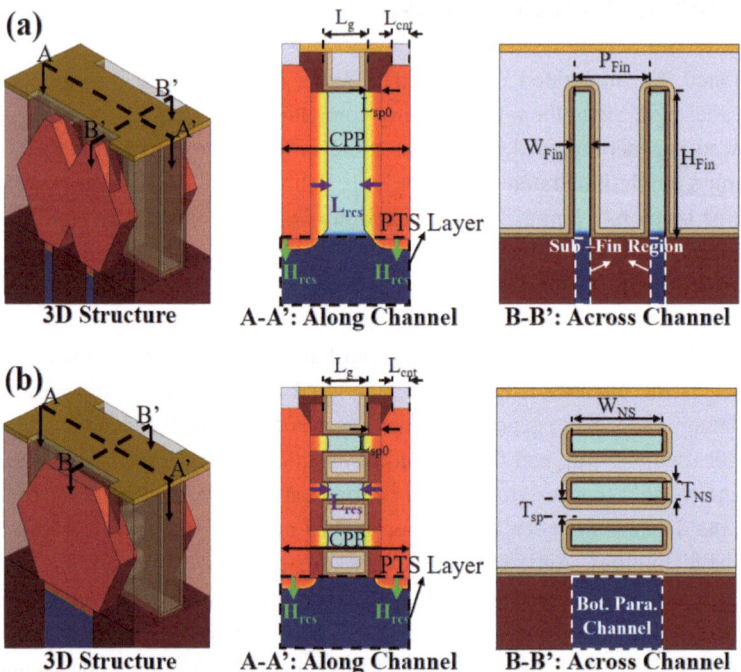

Fig. 1.1 Three-dimensional schematic views of the ideal device and cross-sections along and across the channel for a **a** FinFET, and for a **b** nanosheet FET with the same footprint on the substrate. Reprinted from Wang et al. [11] under the Creative Commons Attribution (CC BY) License 4.0

chips, although the law is most widely invoked in the case of microprocessors. In the present case, the important point is that *Moore's Law is driven by economics, not the technology*. To understand how this continued growth has occurred, it is necessary to realize that economics drives the chip industry (which is not different from most other industries in that fact). Simply put, the cost of a chip, after fabrication, has remained nearly constant over the decades of progress. This cost can be expressed in terms of size, or the area occupied by the silicon chip, which amounts of a few tens of dollars per square centimeter. This means, functionally, that if more processing components are put on a single chip, the cost will be lower for each of those components. For the end user, this means the multi-million dollar Cray computers of the early 1970s will be built for less than a dollar today. That is the nature of progress that drives nanotechnology in our world.

There is a simple reason why this is true. Chips are designed in a rectilinear layout, notably because they can be moved in only two dimensions during fabrication. Thus, the layout of the transistors, or processing units, follows a grid much like the layout of a modern city. Generally, from one generation to the next, a dimension of the transistor is often reduced by $\sqrt{2}$, which means that the area of silicon occupied by this unit is reduced by a factor of 2. If the chip maintains an area of about 1 square centimeter, then this length reduction leads to the ability to put twice as many processing units on the chip. Each now costs half as much. In the earlier days, the chips were not as big as this, so at the beginning Moore noted that progress followed from three factors: (1) the reduction in the linear dimension, (2) an increase in the size of the chip, and (3) circuit cleverness, by which the designer learned to make a processing unit with fewer transistors. The latter two factors often led to an increase in a factor of 4, rather than 2, so that progress was more rapid than expected.

A major challenge in chip design is the interconnection of all these transistors and processing units. Although these processing units are laid out in a planar manner on the chip surface, there are often 10 or more layers of metal that run in the two possible directions and cross one another, much like a 10-level highway interchange. Connections from these lines down to the actual transistors, or processing units, must be by vertical interconnects, referred to "vias." The multiple levels are needed for power distribution, control lines, signal lines, and so on (we will see later that quantum computers are subject to this same problem of interconnects). A second problem is that while the down-sizing reduces the power per transistor, the overall power required by the chip goes up. This has led to several redesigns to accomplish lower power logic over the decades of Moore's law.

Through the evolution of the chip, many significant changes in materials and design of the transistor have occurred. This has included the introduction of strain in the device, both compressive and tensile, that is used to improve the mobility, and thus the current, in the device. Alternate insulators, with higher dielectric constant, have been introduced to avoid tunneling through very thin oxides in the down-scaled devices. Less than a decade ago, the transition to vertical transistors in order to control the off-current was made with the introduction of the FinFET (Fig. 1.1) [12]. Then, the need to have wider current-carrying channels, in order to increase the total current, started driving the industry toward nanosheet transistors [13]. Thus, continued evolution has not been achieved merely by scaling, but by extensive research in materials and device design, and this still is occurring.

Current State of Chip Technology. Advances in circuit cleverness (item 3 in Moore's law above) has led to further innovations in chip design. Almost as soon as the microprocessor was introduced, graphical games appeared on the scene. While these were large consoles in the beginning (leading to the creation of video arcades), the evolution of graphics processing cards led to the ability to put gaming into the home computer (never, of course, on a business computer!). This, in turn, led to the development of the graphical processing unit (gpu) as a stand-alone separate microchip. Games are largely dependent upon image processing to update the player's screen during game play, and for the computation of the update. "Gaming" computers place a premium on the gpu, and the resultant speed and precision of the updates as the player moves through the game (which necessitates the updates). Modern versions of the gpu, however, are capable of many standard operations such as translations, rotations, shading, and even Fourier transforms. These are largely matrix operations. But, these operations also are standard operations in Hamiltonian dynamics in classical mechanics and mathematics. Even a high-order differential equation can be separated into a set of first-order equations and this leads to matrix operations. As a result, most high-end microprocessors today utilize embedded gpus that are utilized in normal operations, and especially in the neurally-inspired processors.

An example of this scientific application is the modern supercomputer, which normally gains it massive speed from extensive parallelization of the processing. Supercomputers are evaluated by a set of standard operations with the results tabulated at sites such as Top500.org [14]. As of June 2024, the top/fastest supercomputer was the exascale Frontier machine, which is a Cray EX235, with a performance of almost 1680 PetaFLOPS (a FLOPS is a FLoating point Operation Per Second; this speed is just over 1.6×10^{18} FLOPS, using 64 bit floating point numbers). This machine contains 8,699,9044 cores, or nodes. Each node contains an AMD microprocessor (the central processing unit, or cpu) and 4 AMD gpus. This operational capability is more than a factor of 3 greater than the next supercomputer on the list, an IBM supercomputer, and is the first machine to reach the exascale level (more than 10^{18} FLOPS).

But, most people don't use supercomputers, they use laptops. For comparison, the Apple M1 Pro chip (late 2022), found in desktops and the Macbook Pro laptop, uses 5 nm technology and has 10 cores in its cpu, but it also has a gpu with up to 16 cores, along with a 16-core neural engine. The gpu alone is said to have a throughput of up to some 5.2 teraFLOPS (5.2×10^{12} FLOPS). While this is a mere shadow of the performance of the top supercomputers, it is far beyond early supercomputers, where the early Cray I (1975) could manage only 1.6×10^8 FLOPS. The top supercomputer in the year 2000 managed only just over 7.2 teraFLOPS, which says that the modern laptop is just a little slower than the top supercomputer of that year.

If it seems hard to get one's head around these numbers, think about Moore's law, and the size of these devices. The top supercomputer at the end of 2000 required some 200 cabinets and occupied 12,000 ft^2. It also consumed some 6 megawatts of power, including its cooling needs, and weighed on the order of 100 tons. This computing power was available to a small set of scientists who needed top end computing power. Today, this computing power is contained in a microprocessor that occupies about 2.5 cm^2 and is specified to consume 30 watts. This powerful computer can be had in desktops and laptops by anybody who desires them. And, more powerful laptops and desktops are arriving regularly.

1.2   **Classical Computation**

From the time of Turing, the development of computing systems followed a path in which the builders imagined they were creating a Turing machine [2]. Turing called his central unit the automatic machine, or *a*-machine, and finally just a machine. Turing considered that his machine would be composed of a head, which he termed the automaton, and a finite-dimensional tape (this tape forms both the memory and the registers in which operations are performed). The automaton would also make decisions about writing and/or advancing the tape. The tape served as the medium upon which symbols were read, written, and stored. The automaton and the tape today could be referred as the combination of the central processing unit (cpu) and the memory. Two types of symbols could be written on the tape, normal binary numbers, our usual 0 and 1 symbols, and special symbols that were non-binary. These special symbols were needed in the various operations, while the binary numbers were essentially the data. One such number was a decimal point, which when placed in front of the next number told the user that the computation was finished.

Importantly, Turing's *a*-machine could remember prior symbols read from the tape, that is symbols that were stored elsewhere than the current position of the tape. This memory meant that there existed a storage register that likely was part of the automaton. One important point was that no blanks could exist on the tape until the end of the computation was reached. A final blank would cause the machine to stop. Turing went to great length to describe the full set of numerical processes that the machine used to compute a number. This enumeration is not of any great interest in the present discussion. But, he also listed a number of criteria which the machine must meet, and these are of interest.

Turing's reason for developing this machine was to address the *Entscheidungsproblem*. This mathematical problem is one of 23 unsolved problems enumerated by David Hilbert at a mathematical conference in Paris in 1900 [15]. The *Entscheidungsproblem* was the 10th on the list, and asked whether a researcher could prove that a statement was either true or false, using only logical axioms. Turing assumed a common variant as the basis for whether or not a number is computable, if one is given set of initial conditions [16, 17]. He assumed that if the machine did not stop, then obviously the number was not computable. Thus, it became clear that one could not predict whether or not the number was computable in advance of the calculation. Only after the computation would one have the answer.

In considering the operation of the machine, Turing laid down some principles of operation. The key principles for the Turing machine (and therefore for any computer considered to be an implementation of such a machine) are that the sequences must be "circle-free" and that there is a definite a stop/end state. Recall that a blank space on the tape signals that the computation has ended. The answer is given just before the blank. This requirement, by itself, means that the computation must proceed to a logical conclusion that gives the numerical answer. Turing also requires that a force is applied to guide the machine toward this stop/end state. These latter two requirement both describe an arrow of time which means that time reversal symmetry is broken, and the machine operates in a non-equilibrium state. It has been recognized for quite some time that the application of a force leads to a phase transition (that breaks time-reversal symmetry) and requires energy dissipation [18].

Many other reasons exist for Turing's machine to be irreversible, both physically and logically. The tape must store information and keep it from thermalizing, which would randomize all information on the tape. In order to maintain the information, the tape must exist in a non-equilibrium, ordered state [19]. Symbols placed on the tape must follow certain logical rules, and these must be unaffected by random thermal noise, thus the need for the non-equilibrium state. Turing also requires the computation to proceed to a logical conclusion. The existence of the arrow of time also means that the force is accompanied by dissipation.

Bits. Turing desired that his numbers be expressed as binary numbers, perhaps because of the usage of 0 and 1 in logic. This form of logic was developed by George Boole to describe the truth and falsity of philosophical arguments [20]. Today, the use of binary numbers is generally referred to as Boolean logic. The binary number scheme was adopted for use in most early machines, such as Zuse's Z3 [21] in Germany and the Atanasoff-Berry computer [22] in the United States. While there were some deviations, this approach has become largely mainstream. In binary computing, the bit is the essential element. It is useful in modern electronics, because its two states, either 0 or 1, can be represented by the two states of a single transistor, which is either off or on. The advantage of this is the restoration of any noise error at each transistor, with output being either the bias voltage (the off state) or ground (the on state). This provides natural error correction throughout the computation.

The bit provides the basic element of the binary word, since it can be represented mathematically as existing in only two states. A representation of this may be expressed as

$$|psi\rangle = a|0\rangle + b|1\rangle, \tag{1.1}$$

where here the brackets represent a state and a and b satisfy

$$a + b = 1, \{a, b\} \subset (0,1). \tag{1.2}$$

These equations mean that a, b can have only the values 0 or 1, and their sum is 1. In (1.1), the term on the left is the state of the bit, and the resultant equations and definitions imply that this bit must be either 0 or 1. This form is introduced here, because it will be used later for the qubit (or quantum bit), and it is easier to explain the difference with these equations.

Inside the computer, a particular state of the various bits defines a machine state. As a computation step is processed, the state of the machine changes. The change in the machine at each step may be expressed as a state transition matrix. The actual state may be defined in a number of ways. For example, if there are only 3 bits in a very simple machine, then there are 8 states possible (2^3). This can be expressed as a 3-bit word or by a 8-bit word. In the former the word is composed of 3 binary terms that are either 0 or 1, and this will be called the bit matrix. In the latter, the word has a single 1, with the other bits being 0, so that the single 1 denotes in which of the 8 states the machine lies. This form will be called the state matrix. The state transition matrix is more easily understood in this latter form, which will be used below in several discussions. The state transition matrix indicates the action that is taken when a control signal is applied (this control signal often selects the proper state transition matrix among the various possibilities). Hence, an algorithm is a map that guides the machine through desired states to produce (hopefully) an answer. While this is an extremely simple model, it provides enough detail to discuss physical limits to the computing process.

When the bits are combined into a computer, it is important to understand that this system is a physical system that is subject to the laws of physics. These laws describe the dynamics of the system, principally through the various degrees of freedom that exist in the machine. There are many parts of this system. For example, a large number of bits compose the information-bearing (IB) portion of the machine. This may be the memory, the registers, and the arithmetic logic unit in a typical machine. On the other hand, there may be many more bits that are non-information bearing (NIB), and these are certainly various control bits and bits that enable, for example, a memory read operation. But, these also include the bits necessary to describe the physical state of the system. This is depicted in Fig. 1.2, where the IB and NIB are defined. The figure also indicates the input of power and information, and the output of heat and (hopefully) new information. Control of the information is accomplished by the program and the algorithm. Control of the power and heat is through design of the physical system in which the computing bits are embedded. The entire system is then embedded in the world and connected to the environment provided by this connection. Thus, the IB bits are only a small part of the computing system, and are a miniscule fraction of the total physical states in the system. While the IB bits are macroscopic objects which may be read and changed appropriately, they are defined at the microscopic level by the various physical processes [23]. For example, an IB bit may be the charge on a memory capacitor, but the physical nature of this capacitor may entail an extremely large number of microstate variables. These microstates may be represented by NIB bits.

Irreversibility of Computing. Thus, we understand from Turing's description of the requirements on his machine that the system is irreversible. This is inherent in the arrow of time and the resulting dissipation (heat generation). On the other hand, there is a question as to whether reversible logic can provide computational abilities, supposedly with less heat dissipation. This question came to the forefront with the advent of concepts for quantum computation [3]. Textbook quantum mechanics deals with closed systems for which the Schrödinger equation is time reversible. Thus, early on, it was initially thought that

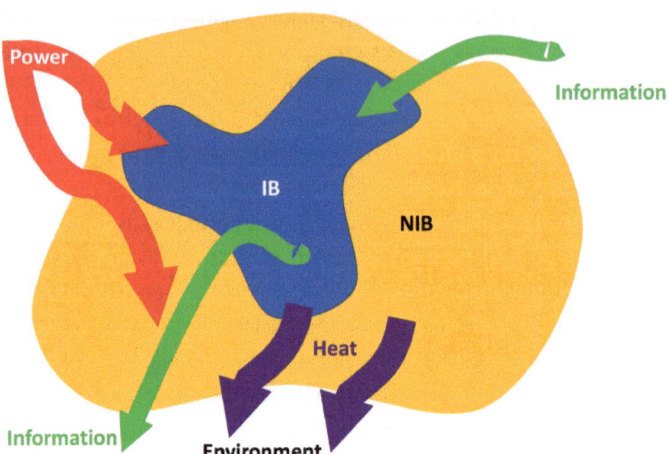

Fig. 1.2 The information-bearing bits are embedded in a much larger sea of non-information bearing bits, the latter of which include a representation of the microstates of the computing system. Both are connected to the environment, where heat must eventually be deposited

quantum computers could be reversible. However, it is clear that such machines would not be Turing machines. Further advances eventually demonstrated the need for dissipation even in quantum computers [18, 22]. But, this does not always preclude reversible logic, so that the question of whether or not such logic can be useful. The answer is no, but this still needs to be shown.

The prototypical reversible gate is the Fredkin gate [24]. It possesses an equal number of inputs and outputs, so that one can infer the inputs from the outputs. The mapping from one to the other is 1:1. But, it is well known that the Fredkin gate can be implemented with normal CMOS circuitry. This means that it is possible that reversible logic can coexist with dissipation. So the question of reversible logic remains of interest.

In the previous sub-section, the bit matrix and the state matrix were introduced. Accordingly, a conversion matrix exists to transform one into the other. This matrix arises from the topology of the computing circuit itself, and is known in circuit theory to be related to the tie-set matrix [25]. Consequently, the description of one of these two matrices (the bit and state matrices) is fully equivalent to the other. Thus, let us concentrate on the state matrix. The state transition matrix is an 8×8 matrix. Reversible logic requires that each state have a unique successor and a unique predecessor. That is, "fan-in" and "fan-out," for example, are forbidden actions, as each reduces (or increases) the phase space occupied by the dynamical variables of the bits. Reversibility precludes these increases or decreases. Thus, the existence of a unique successor means that each row of the state transition matrix has a single 1, and each column of this matrix also has a single 1. The need for the inverse of the matrix means that it is full rank so that every row and column has a single 1. Now, it turns out that this matrix is a characteristic matrix for the cyclic permutation group, which means that each state lies on one or more rings. Thus, the state propagates around these rings, or loops, with no clear stop state. This violates two of Turing's requirements for his computing machine. He does not allow rings, or loops, and requires a definite stop state. Thus, the conclusion must be that a machine with only reversible logic cannot emulate a Turing machine.

1.3   Quantum Technology

Interest in the quantum world arose outside of science with the first thoughts about quantum computers [4]. As this interest grew, it became evident that the basic principles of quantum mechanics could be utilized in fields other than computing. Notable among these is communications theory, where encrypted communications are of interest to those sending the messages as well as others, beyond the recipient, who would like access to the content of the message. Reading other people's mail, and keeping people from doing this, has been of interest as long as governments have existed [26]. Quantum communications is just the latest approach to keeping communications secure, and quantum computing has been suggested to help make them less secure. It was actually found earlier that quantum technology could be useful in the world of measurements, so that quantum sensing came to the forefront. This is because superconducting qubits usually rely upon the superconducting quantum interference device (SQUID) as an important part of the qubit. In fact, the SQUID has been used to detect low levels of magnetism around the world for more than half a century. So, quantum sensing is actually an older field than the other parts of quantum information. In this section, the general outline of these fields will be discussed, although detailed applications are left to Chap. 2.

Qubits. The difference between bits and qubits is the difference between digital and analog system. To begin, the qubit still satisfies (1.1), but (1.2) is dramatically changed. The relationship between a and b now becomes

$$|a|^2 + |b|^2 = 1. \tag{1.3}$$

Here, it is seen that the amplitude of the qubit remains unchanged from the bit, but the two coefficients are now continuous complex variables, subject only to keeping the magnitude unity. Thus, the qubit is actually an analog variable, or, more properly, the qubit has unit amplitude but a continuous analog phase. Thus, operations on the qubit are mainly phase rotations; these will be discussed more fully in Chap. 2. But, by itself, the qubit offers little new to the computing world, as can be seen from one of the main quantum algorithms–the quantum Fourier transform. Having said that, it must be pointed out that qubits cannot be made from classical systems; in order to achieve their power, they must be made from quantum systems.

There is a belief that digital quantum qubits can exist, different from analog type qubits. These gates are said to operate by rotating and shifting the states between different superpositions of the one and zero states, as well as providing different entangled states [27]. But, as will be seen in Chap. 2, this is exactly what variations in the complex numbers a and b do with the qubit of (1.3). Perhaps this is a subtle difference in viewpoint, but one can regard such rotations as analog variations of the phase of the wave function.

In a Fourier transform, which is one of the basic elements in communications and science, a data set arises from sampling a time varying signal. The signal can then be decomposed into its various frequency components by the Fourier transform. If the signal to be converted is digital samples, then the transform is referred to as the discrete Fourier transform, or DFT. If there are N time samples, then Fourier coefficients can be obtained for N different frequencies. While the DFT is relatively slow, a very fast version was obtained as the Fast Fourier Transform [28]. This resulted in a reduction of computing time of $O(N^2)$. The secret was to use the binary system so that each frequency differed from its neighbor by a factor of 2. That is, the FFT is computed from the summation

$$F_k = \sum_{n=0}^{N-1} t_n exp(2\pi i k n/N), k = 0,1, 2, \ldots N - 1. \tag{1.4}$$

Here, t_n are the N time samples of the signal, and F_k are the complex amplitudes of the Fourier components.

In describing the quantum Fourier transform (QFT), it is often compared to the DFT, when it should be compared with the FFT in discussing any speedup from the quantum version [29]. In fact, the algorithm usually cited for the QFT is only a variant of the FFT. The usefulness of the QFT lies not in the qubits themselves, but rather in what can be done with qubits that cannot be done with bits, and that is *entanglement*. A glance at (1.4) tells us that the summation process must be repeated N times, once for each value of k. The power of quantum computation lies in entangling all N sums into the qubits all at the same time. Entanglement is the secret power of quantum mechanics [30]. This one shot is often not enough, because quantum mechanics is a probabilistic theory. Thus, one needs to perform the computation a few times to assure the correct answer, but this is less than N, so that the QFT speeds up the process.

Entanglement. Since the beginning of quantum mechanics, a debate raged over the interpretation, or the philosophy, that governed this new science [8]. This had led to a series of arguments between Einstein and Bohr that dominated the 1927 Solvay conference dedicated to this new science [31]. In 1935, the debate continued to be an issue. Early that year, Einstein, Podolsky, and Rosen (EPR) submitted a paper questioning the idea that quantum mechanics was a complete theory [32]. Here, they considered two interacting particles and their behavior after the interaction. They questioned how measuring one of the particles and inferring the other in Bohr's manner required a "spooky action at a distance" [32]. From further arguments, they asserted that quantum mechanics must be an incomplete theory. Bohr responded, but did not really answer the questions [33]; in fact, the next year Bohr would basically admit that the theory might "…need a more comprehensive generalization…" [34]. It remained for Schrödinger to point out the missed point of the EPR argument, and this was entanglement [30].

When two particles interact quantum mechanically, they can no longer be considered as separate entities. Rather, instead of two separate wave functions prior to the interaction, they now become described by a single two-particle wave function; that is, they are now entangled. While this latter wave function certainly has probability peaks at the locations of the two particles, it exists over much of the distance between the two particles. Now, a measurement of one is actually a measurement of both, since it applies to this single two-particle wave function. Entanglement assures that the properties of the two particles are now correlated in a manner that determining the property of one certainly gives information on the same property of the other. This means that there is no need for Einstein's "spooky action at a distance." Schrödinger clearly felt that entanglement was the most important property of quantum mechanics. This would be made more explicit in a later discussion from Bohm [35].

Entanglement provides the magic that, in turn, provides quantum computing with its advantage. If all values of k, in (1.4), can be entangled into the qubits, then the summation will yield the Fourier coefficients in an entangled form. It remains only necessary to de-entangle the result to achieve the desired QFT. Now, preserving the process, as well as the result, requires minimum decoherence, as the role of decoherence is to destroy entanglement. As will be seen in the next chapter, major efforts for error correction, and to minimize dissipation/decoherence, are the norm in quantum computing.

Quantum Computing Today. Generally, quantum computing moved beyond the realm of mathematical curiosity when it was suggested that arrays of weakly interacting qubits could be entangled by the use of controlled electromagnetic pulses [36]. This shifted the development to algorithms that would arrange for the required superpositions/entanglements. Perhaps the first such algorithm was the quantum Fourier transform, discussed above, as a method of factoring large numbers [37]. A second algorithm appeared as a method of speeding up database searches [38]. Many other suggestions for quantum algorithms appeared, but usually only stimulated the development of new classical approaches that were just as fast [39]. This, however, did not slow the development of real systems for quantum computation.

The first actual quantum computer seems to be the D-Wave system in 2010. This computer differed from the above discussions on Turing machines as it was designed to work on minimization, or optimization, problems. In such problems, an "energy" functional is created, and this "energy" is then minimized by a process known as artificial annealing. The size of this machine seems to have grown to 5000 qubits in the 2023 Advantage machine that was used in a spin-glass problem [40]. In addition to the annealing machine, the company is expanding into traditional gate-based machines. In both cases, the machines are based on Josephson junctions (to be discussed in Chap. 2), which are phase coherent tunnel junctions using superconducting materials. Such junctions yield macroscopic quantum behavior [41]. To date, the company has delivered more than a dozen machines to various companies and government laboratories.

Today, perhaps the best-known name in quantum computation remains IBM, likely because of their long presence in classical computing. In 2022, they introduced a 433-qubit processor, which at the time was said to be the largest quantum computer, and this may be true for a machine aimed to be universal in the Turing sense. At the same time, they also presented a roadmap which predicts 1100 qubits in 2023, and something like a million qubits within another decade. Certainly, IBM has led in establishing a user network that brings together a great many universities, government labs, and industries into their user network. The company attached their early implementations of quantum processors to one of their mainframe machines as a co-processor, and this has allowed the development of many new approaches in all parts of quantum information, by allowing users to focus on software generation and new circuit configurations for particular applications [27, 42]. Following today's supercomputer methods, new quantum processors will actually provide parallel processors to enhance the overall performance. Other companies, such as Rigetti and Google, use this superconducting technology.

A new company in the quantum world is Quantinuum, created from Honeywell and Cambridge Quantum Computing. Their approach is based upon the concept of trapped ions. The qubit is formed from states of the trapped ion, either the ground and an excited state, or a pair of hyperfine separated states in the ground state itself. The latter mode is used in the Quantinuum machine with the hyperfine states existing in the ground state of Yb^+ ions [43]. The traps are formed by radio frequency fields that provide a radial energy minimum in which the ion sits. Lasers are used to excite the qubit to one or the other of the two levels (the ion is still basically a two-level system). In this system, the ions are mobile, moving along a series of traps much as electrons in a charge-coupled device (CCD, an electron shuttling device used in many imaging arrays). There are other trapped-ion approaches; one of these is IonQ, which has a quantum processor accessible on many cloud systems. The structure of the IonQ machine is very similar to that of the Quantinuum machine, using radio frequency and constant voltages to create the Yb trap, and stabilized diode lasers to address the qubits [44], which also use the hyperfine levels as the two states [45]. The IonQ machine does not appear to shuttle the qubits themselves as done in the Quantinuum machine. One advantage of the trapped ion implementations is that they are said to be extendable to room temperature, which would be a major advantage [46]. Other companies using trapped ions are Alpine QT, Eleqtron, and Alpine QT.

A slightly different approach has been followed by Pasqal, which uses neutral atoms as the qubits. In this approach the neutral atoms are trapped on an optical lattice as an array. The individual qubits still use the hyperfine levels of the ground state, but other levels can be used for various processes, such as cooling the atoms [47]. Highly excited Rydberg states (well separated from the ground state) are used for qubit processing. In this sense, these ions move beyond the simple two-level system. But, these neutral atoms still remain cold atoms held at low temperature [48].

The Microsoft approach follows a totally different effort in physics, using topological states of matter as the prototypical quantum computer. These qubits interact by moving around one another, in a process termed "braiding" [49]. The interaction is governed by a rather obscure mathematics known as "knot theory," and, yes, it describes how you tie your shoes [50]. In the Microsoft approach, the individual "particles" to be moved around are taken to be Majorana fermions. The latter is a particular state of zero energy in which the particle is its own anti-particle. Majorana pointed out that this type of particle was missed by Dirac in his relativistic quantum mechanics, and its properties were described by him in the late 1930s [51]. The basic approach uses an Al superconductor adjacent to an InAs semiconductor. InAs is a narrow gap material with a large spin–orbit interaction, and the adjacent Al superconductor induces some superconducting behavior in the semiconductor. Near this junction, conditions are thought to be right for creating Majorana fermions [52]. Surprisingly, Quantinuum has used their trapped ion device to produce the necessary non-Abelian anyons (similar to the Majorana fermions) and braid them topologically [53]. The non-Abelian Anyon is a quantum particle first conceived by Wilczek [54]. The use of topological states has the advantage that they should be immune to noise, and thus require much less error correction.

Intel continues to pursue spin qubits based upon Si quantum dots, following early collaboration with the groups at the University of New South Wales in Australia and a current collaboration with QuTech (founded from research at TU Delft in Holland). They have recently released a qubit chip using their advance Si processing capabilities and demonstrated operation at temperatures above the normal dilution refrigerator operating at milliKelvin temperature [55]. A different approach, using Si dangling bonds is being pursued by Quantum Silicon.

Both Xanadu and QCI are pursuing quantum computing that depends upon photonic techniques. The latter, QCI, actually uses an architecture and approach quite similar to DWave's optimization procedure, but using the photonic approach after their acquisition of QPhotonic. These approaches, and some others beyond the above discussion, will appear in Chap. 2.

1.4   Quantum Communications

The goal of quantum communications is to use the properties of quantum mechanics, specifically qubits, to provide efficient, and secret, transmission of information whether to another person, to the bank, or to a quantum computer. It seems that a week doesn't go by without news of another hack of someone's supposedly safe file system. The hope is that by using quantum techniques, this can be avoided in the future. Certainly, most communication is encrypted, either routinely by your cell phone, or on purpose using a variety of efficient encryption schemes. But, as is well known, there also are a variety of techniques to break various encryption methods. Secret communications have existed for as long as people have communicated, and methods of breaking into these secrets have existed just as long. Perhaps the most famous is the Germans reliance upon their Enigma machines in World War II, even though the allies were reading their mail almost faster than the intended recipient, thanks to the Polish security group getting their hands on one of the machines well before the war [26].

There are a multitude of so-called channels in the communications world. The most immediate is over-the-air communications such as that with cell phones. Then, there are cables, either RF such as microwaves, or fiber optic versions using lasers. Much communications involve satellites. But, cables can be tapped, and nation states have been listening to over-the-air and satellite communications for decades. This all means that cryptography is not just an interesting science; it is the backbone of modern communications. Quantum mechanics offers new hopes for secure communications. The properties of an entangled qubit means that any attempt to read it will cause the qubit to collapse into either a 1 or a 0. When the recipient sees this classical state, he/she knows there has been a hacking attempt. However, with any encryption technique, the recipient has to know the key necessary to decrypt the signal readily and recover the information. For example, the four-digit code a user has on his/her bank card is used by the bank to aid in decrypting the signals from an ATM. Similarly, in communications the recipient needs to have a *key* that aids in the decryption of the message. Thus, secure quantum communications needs three important factors: (1) an encryption technique, often referred to as the protocol, (2) a method of sending the key to the recipient, known as key distribution, and (3) a repeater scheme that can amplify the signal while preserving the encryption and the entanglement. These are the elements of the communications channel.

Combining satellites, cables, and other approaches all can contribute to a communications network. Transitioning to quantum information networks is a goal shared now worldwide [56]. There is an important aspect to these networks that needs to be mentioned, and that is time synchronization [57]. For example, in a famous physics experiment, a pair of entangled photons were sent to two different sites, in order to try to evaluate the correlations in measurement. But, one has to be sure that the two entangled photons are simultaneously measured, and this requires a synchronization signal. Similarly, modern communications, even quantum communications, benefits from time synchronization. Early attempts for secure quantum communication used the ideas of teleportation: the state of a qubit could be transmitted to create an identical entangled cubit somewhere in the distance while destroying the local qubit. Hence, the idea arose that the present qubit had been teleported [58]. However, the development of encryption algorithms today are based upon the need to have quantum-resistant algorithms, meaning that they cannot be broken by quantum techniques. These are sometimes called post-quantum cryptography [59]. As mentioned, this requires a method for key distribution, and this should not be dependent upon the device technology that is implemented

[60]. There is at least one view that currently this can be done efficiently via optical channels [61].

Similarly, there has been progress in the creation of repeaters, that are necessary to enable truly long-distance communication in cable networks (it would be hard to implement intermediate repeaters for satellite communications). Repeaters for telecom wavelengths have been recently achieved using trapped ions [62]. Other approaches are certainly available. We will return to this topic in Chap. 2.

1.5 Quantum Sensing

Sensing is at its heart the world of measurement and is often required in unfavorable environments. Classically, noise has always been a problem in measuring very sensitive signals. Many decades ago, this noise limit was surpassed by modulating the incoming signal and using this modulation to measure signals below the noise level of the receiver/sensor [63]. Since then, the measurement of sensitive signals has naturally progressed. Today, it is hoped that quantum sensors, that exploit the properties of light and atoms (such as in qubits), can be utilized to realize extremely sensitive detectors/sensors. Yet, as pointed out above, quantum sensing is much older than the other two parts of quantum information. The SQUID device has been used as a sensitive detector of magnetic fields for more than half a century. Whether it is looking for magnetic anomalies in the ocean, studying nuclear magnetic resonance in chemistry and physics, or using magnetic resonance imaging in medicine, the SQUID is likely to be present. This follows from the use of superconducting magnets in everyday life that provides a natural environment for the SQUID.

Of course, merely developing sensors is not the end product; a method of properly using these sensors also has to be developed. That is, the usefulness of a sensor itself must be coupled to the proper system optimized to access the proper properties of the sensor [64]. For example, it is known that a key element in quantum sensing is that the final step in the measurement, of extracting the information from the quantum sensor, is itself not free from noise, whether this noise be classical or quantum in origin. This leads to a need to develop quantum algorithms that aid the ability of the system to achieve a useful, or even valid, estimate of the actual signal being sensed [65].

The realm of quantum sensors spans the entire field of appropriate science for qubits, and the SQUID will appear here in superconducting qubits, discussed in Chap. 3. In detail, one needs the creation of quantum sources, usually entangled qubits, and then quantum measurements that, together, allow one to do better than normal classical schemes. Generally, the need is for an ability to achieve coherent control of the quantum system while not increasing the noise [66]. It appears that many approaches are using optical systems in quantum sensing, as generation of entangled photons seems to be a more advanced technology [67, 68], although spin qubits are certainly also being pursued [69], as are all of the previously mentioned types of qubits in quantum computing. The optical techniques, in particular, look toward using entangled photons and/or squeezed (and therefore nonclassical) states. These will be described in more detail in the following chapters.

References

1. Menabrea, L. F.: Sketch of the Analytical Engine invented by Charles Babbage. Tr. by Lovelace, A. A. Taylor's Scientific Memoirs 3, 666 (1843).
2. Turing, A. M.: On Computable Numbers with an Application to the Entscheidungsproblem. Proc. London Math. Soc. 42, 230 (1937).
3. Benioff, P. A.: Quantum Mechanical Hamiltonian Models of Discrete Processes That Erase Their Own Histories: Applications to Turing Machines. Int. J. Theor. Phys. 21 177 (1982).
4. Deutsch, D.: Quantum Theory, the Church-Turing Principle and the Universal Quantum Computer. Proc. Roy. Soc. London A 400, 97 (1985).
5. Kilby, J. S.: Miniaturized Electronic Circuits. U. S. Patent 3,138,743, filed 6 February 1959, issued 23 June 1964.
6. Evans, C.: The Micromillenium. Washington Square Press, New York, 1982.
7. Moore, G. E.: Cramming more Components onto Integrated Circuits. Electronics 38(8), 114 (1965).
8. Ferry, D. K.: The Copenhagen Conspiracy. Pan Stanford, Singapore, 2019.
9. Oriols, X., Ferry, D. K.: Why Engineers are Right to Avoid the Quantum Reality Offered by the Orthodox Theory. Proc, IEEE 109, 955 (2021).
10. Dennard, R. H., Gaenslen, F. H., Yu, H.-N., Rideout, V. L., et al.: Design of Ion Implanted MOSFETs with very Small Dimensions. IEEE Sol.-State Circ. 9, 256 (1974).
11. Wang, D., Sun, X., Liu, T., Chen, K., et al..: Investigation of Source-Drain Recess Engineering and Its Impacts on FinFET and GAA Nanosheet FET at the 5 nm Node. Electron. 12, 770 (2023).
12. Hisamoto, D., Lee, W.-C., Kedzierski, J., Anderson, et al.: A Folded-Channel MOSFET for deep-submcron era. IEEE Electron Devices Meeting Digest, p. 1032 (1998).
13. Jang, D., Yakimets, D., Enemen, G., Schuddinck, P., et al.: Device Exploration of NanoSheet Transistors for sub-7 nm Technology Node. IEEE Trans. Elec-tron Dev. 64, 2707 (2017).
14. https:www.top500.org.
15. Hilbert, D., Ackermann, W.: Grundzüge der theoretischen logic. (Berlin: Springer-Verlag, 1928) Tr. by authors as Principles of Mathematical Logic (AMS Chelsea, Providence, RI, 1950).
16. Gödel, K.: Die Vollständigkeit der Axiome des logischen Funktionenkalküls. Monatshefte Math. Phys. 38,173 (1931).
17. Church, A.: An unsolvable problem of elementary number theory. Am. J. Math. 58, 345 (1936).
18. Prigogine, I.: Dynamical Roots of Time Symmetry Breaking. Phil. Trans. Royal Soc. A Math. Engr. Phys. Sci. 360 299 (2002).
19. Porod, W., Grondon, R. O., Ferry, D. K., Porod, G.: Dissipation in Computation. Phys. Rev. Lett. 52 232 (1984).
20. Boole, G.: An Investigation of the Laws of Thought. (Walton and Maberly, London, 1854).
21. Rojas, R.: Konrad Zuse's Legacy: The Architecture of the Z1 and Z3. IEEE Annals Hist. Comp. 19, 5 (1995).
22. Mollenhoff, C. R.: Atanasoff: Forgotten Father of the Computer. (Iowa State Univ. Press, Ames, 1988).

23. Ferry, D. K., Porod, W.: Dissipation and Irreversibility in Computing. Mater. Res. Express 10, 083001 (2023).
24. Fredkin, E., Toffoli, T.: Conservative Logic. Int. J. Theor. Phys. 21, 219 (1982).
25. Guillemin, E. A.: Introductory Circuit Theory. (Wiley, New York, 1953).
26. Kahn, D.: The Code Breakers: A Comprehensive History of Secret Communications from Ancient Times to the Inernet. (Schribner, New York, 1966).
27. Rietsche, R., Dremel, C., Bosch, S., Steinacker, L., et al.: Quantum Computing. Electron. Markets 32, 2525 (2022).
28. Cooley, J. W., Tukey, J. W.: An Algorithm for the Machine Calculation of Complex Fourier Series. Math. Comp. 19, 197 (1965).
29. Nielsen, M. A., Chuang, I. L.: Quantum Computation and Quantum Information. (Cambridge, Cambridge Univ. Press, 2000).
30. Schrödinger, E.: Die gegenwartige Situation in der Quantenmechanik. Naturwiss. 23, 803, 827, 844 (1935); Tr. Trimmer, J. D.: The Present Situation in Quantum Mechanics: A Translation of Schrödinger's "Cat Paradox" Paper. Proc. Am. Phil. Soc. 124, 323 (1980).
31. Bacciagaluppi, G., Valentini, A.: Quantum Theory at the Crossroads. (Cam-bridge, Cambridge Univ. Press, 2013).
32. Einstein, A., Podolsky, B., Rosen, N.: Can Quantum-Mechanical Description of Physical Reality be Considered Complete? Phys. Rev. 47, 777 (1935).
33. Bohr, N.: Quantum Mechanics and Physical Reality. Nature 136, 65 (1935).
34. Bohr, N.: Causality and Complementarity. Phil. Sci. 4, 289 (1937).
35. Bohm, D., Aharonov, Y.: Discussion of Experimental Proof for the Paradox of Einstein, Rosen, Podolsky. Phys. Rev. 108, 1070 (1957).
36. Lloyd, S.: A Potentially Realizable Quantum Computer. Science 261, 1569 (1993).
37. Shor, P. W.: Algorithms for Quantum Computation: Discrete Logarithms and Factoring. In: Proc. 35th Ann. Symp. Found. Comp. Sci. (IEEE, New York, 1994) pp. 124–134.
38. Grover, L. K.: A Fast Quantum Algorithm for Data Base Search. Proc. Ann. ACM Symp. Theory Comp. (ACM, New York, 1996) pp. 212–219.
39. Tang, E.: A Quantum-Inspired Classical Algorithm for Recommendation Systems. Proc. 51st ACM SIGACT Symp. Theory Comp. (ACM, New York, 2019) pp. 217–228.
40. King, A. D., Raymond, J., Lanting, T., Harris, R., et al.: Quantum Critical Dynamics in a 5000 Qubit Programmable Spin Glass. Nature 617, 61 (2023).
41. Josephson, B. D.: The Discovery of Tunneling Supercurrents. Rev. Mod. Phys. 46, 251 (1974).
42. Alt, R.: On the Potentials of Quantum Computing--An Interview with Heike Riel from IBM Research. Electron. Markets 32, 2537 (2022).
43. Pino, J. M., Dreiling, J. M., Figgatt, C., Gaebler, J. P., et al.: Demonstration of the Trapped-Ion Quantum CCD Architecture. Nature 592, 209 (2021).
44. Allen, S., Kim, J., Moehring, D. L., Monroe, C. R.: Reconfigurable and Programmable Ion Trap Quantum Computer. 2017 IEEE Conf. Rebooting Comp. (IEEE Press, New York, 2017) pp. 1–3.
45. Debnath, S., Linke, N. M., Figgatt, C., Landsman, K. A., et al.: Demonstration of a Small Programmable Quantum Computer with Atomic Qubits. Na-ture 536, 63 (2016).
46. Hensinger, W. K.: Quantum Computer Based on Shuttling Trapped Ions. Nature 592, 190 (2021).
47. Graham, T. M., Song, Y., Scott, J., Poole, C., et al.: Multi-Qubit Entanglement and Algorithms on a Neutral Atom Quantum Computer. Nature 604, 457 (2022).
48. Williams, H. J.: Versatile Neutral Atoms Take on Quantum Circuits. Nature 605, 237 (2022).
49. Rowell, E. C.: Braids, Motions, and Topological Quantum Computing. Arxiv.org:2208.11762v1.

50. Adams, C. C.: The Knot Book: An Introduction to the Mathematical Theory of Knots. (W. H. Freeman, New York, 1994).
51. Majorana, E.: Teoria simmetrica dell'electrone e del positrano. Il Nuovo Cim. 14, 171 (1937).
52. Aghaee, M., Akkala, A., Alam, Z., Ali, R., et al.: InAs-Al Devices Pass the Topological Gap Protocol. Phys. Rev. B 107, 245423 (2023).
53. Iqbal, M., Tantivasadakarn, N., Verresen, R., Campbell, S. L., et al.: Creation of Non-Abelian Topological Order and Anyons on a Trapped Ion Processor. arXiv:2305.03766.
54. Wilczek, F.:Dissassembling Anyons. Phys. Rev. Lett. 69, 132 (1992).
55. Petit, L., Eenink, H. G. J., Russ, M., Lawrie, W. I. L., et al.: Universal Quantum Logic in Hot Silicon Qubits. Nature 580, 355 (2020).
56. de Forges de Parny, L., Alibart, O., Debaud, J., Grassani, S., et al.: Satellite-Based Quantum Information Networks: Use Cases, Architecture, and Roadmap. Commun. Phys. 6, 12 (2023).
57. Nande, S. S., Paul, M., Senk, S., Ulbricht, M., et al.: Quantum Enhanced Time Synchronization for Communication Network. Comp. Networks 229, 109772 (2023).
58. Soares-Pinto, D. O.: Quantum Information Science: From Foundations to New Technologies. Phys. B: Cond. Matter 653, 414510 (2023).
59. Joseph, D., Misoczki, R., Manzano, M., Tricot, J., et al.: Transitioning Organ-izations to Post-Quantum Cryptography. Nature 605, 237 (2022).
60. Zapatero, V., van Leent, T., Arnon-Friedman, R., Liu, W.-Z., et al.: Advanc-es in Device-Independent Quantum Key Distribution. Quantum Inform. 9, 10 (2023).
61. Sax, R., Boaron, A., Boso, G., Atzeni, S., et al.: High-Speed Integrated QKD System. Photonics Res. 11, 1007 (2023).
62. Krutanskiy, V., Canteri, M., Meraner, M., Bate, J., et al.: Telecom Wavelength Quantum Repeater Node Based on a Trapped-Ion Processor. Phys. Rev. Lett. 130, 213601 (2023).
63. Dicke, R. H.: The Measurement of Thermal Radiation at Microwave Frequencies. Rev. Sci. Instrum. 17, 268 (1946).
64. Bongs, K., Bennett, S., Lohmann, A.: Quantum Sensors will Start a Revolution--If we Deploy them Right. Nature 617, 672 (2023).
65. Kurdzialek, S., Demkowicz-Dobrzanski, R.: Measurement Noise Susceptibility in Quantum Estimation. Phys. Rev. Lett. 130, 160802 (2023).
66. Chen, Y., Miao, Z., Yuan, H.: Cooperation between Coherent Control and Noises in Quantum Metrology. Adv. Quantum Techn. 6, 2200165 (2023).
67. Ge, W., Jacobs, K., Suhail Zubairy, M.: The Power of Nonclassical States to Amplify the Precision of Macroscopic Optical Metrology. NPJ Quantum In-for. 9, 5 (2023).
68. Pelayo, J. C., Gietka, K., Busch, T.: Distributed Quantum Sensing with Optical Lattices. Phys. Rev. A 107, 033318 (2023).
69. Yoon, J., Kim, K., Na Y., Lee, D.: Characterization and Correction of the Pulse Width Effects on Quantum Sensing Experiments using Solid-State Spin Qubits. Curr. Appl. Phys. 50, 140 (2023).

Processing in the Quantum World

2

Nearly all of the important aspects of quantum information were introduced in the first chapter. But, this previous discussion probably leaves many still unsure about the various processes and how they are implemented to make quantum circuits. Thus, with this chapter, a rather deep dive into the quantum world will be presented in order to understand many of the topics, such as entanglement, qubit operations, many-body physics, and hopefully some quantum circuits used in computing and the information world. By the end of this chapter, it is hoped that the *how* of quantum information will be fully understood. We begin this discussion with a look into the strange world of entanglement.

2.1   Entanglement

As was mentioned in Chap. 1, Schrödinger called entanglement the most important feature of quantum mechanics [1]. Principally, when two particles, that are normally described with each having its own wave function, interact in quantum mechanics, they become described by a single two-particle wave function. They are now in an entangled state. And, this state will survive until something happens to break up the entanglement, which is said to decohere the wave function (and the particles). It is this entanglement that distinguishes the quantum system from the classical system. This also is the important ingredient in quantum information processing that suggests this processing can be more efficient and more rapid than classical information processing. When two qubits interact, they become entangled, and it is this that is important in quantum computing.

Consider an example proposed by Bohm and Aharonov [2]. A molecule consisting of two atoms, each of which has a spin of 1/2 are considered. Normally, if one considers a

© The Author(s), under exclusive license to Springer Nature Switzerland AG 2025   21
D. K. Ferry, *Quantum Information in the Nanoelectronic World*,
Synthesis Lectures on Engineering, Science, and Technology,
https://doi.org/10.1007/978-3-031-62925-9_2

single electron, the spin is either "up" or "down", which are denoted by $+1/2$ or $-1/2$, respectively. This can be true of atoms as well. Near the end of the nineteenth century, the Dutch physicist Pieter Zeeman discovered that atoms had well distinguished energy levels, but that, when a magnetic field was applied, some of these states would split into two separate levels [3], the so-called Zeeman splitting. It was later discovered that the level itself arose from quantization of the energy and momentum states of the atom. The additional splitting however was suggested as a magnetic field effect on an additional angular momentum term of (at the time) unknown source by Pauli [4]. When a visiting student from Columbia, Ralph Konig, suggested that this could be the electron rotating on its own axis, Pauli was dismissive of the idea. Less than a year later, it was found experimentally that this extra angular momentum was indeed the spin of the electron [5]. Normally, there are two electrons that can sit in a quantum state arising from the quantization of energy and momentum. Now, it was clear that these two electrons had to have opposite spin angular momentum. The Bohm-Aharonov suggestion was to consider this molecule consisting of two atoms, each of which had either an up or down spin. Since the two had to be opposite on the two atoms, the total spin angular momentum was zero. They suggested that one could write the wave function as

$$|\psi\rangle = \frac{1}{\sqrt{2}}\left[|\uparrow\rangle_1|\downarrow\rangle_2 - |\downarrow\rangle_1|\uparrow\rangle_2\right]. \tag{2.1}$$

The subscripts designate either atom 1 or atom 2, while the arrows indicate spin up or spin down. In other words, atom 1 could have either spin up or spin down and atom 2 had the opposite. The wave function of (2.1) is an entangled state of the two atoms. To see this better, let us take a slight diversion.

Hilbert Space. Consider the Fourier series. The various sine and cosine functions are known as a basis set, and the signal to be transformed is now expanded in a series over these basis functions. In the case of the Fourier transform, this is an infinite basis set. For a discrete Fourier transform (DFT) of a signal with N data points in time, there will be only N basis functions in the DFT. The set of basis functions in the Fourier transform are normalized and are othogonal to one another, so that they are said to form a Hilbert space. In the normal Fourier transform, it is an infinite Hilbert space, while in the DFT it is a N-dimensional Hilbert space. In Chap. 1, the qubit was introduced as

$$|\psi\rangle = a|0\rangle + b|1\rangle \tag{2.2}$$

with

$$|a|^2 + |b|^2 = 1. \tag{2.3}$$

Now, there are only two basis functions, the $|0\rangle$ and $|1\rangle$ states. Thus, this forms a two-dimensional Hilbert space, often denoted as H_2.

Consider now the molecule proposed by Bohm and Aharonov. Were each atom a qubit, one way of defining it might be to write the wave function as

$$|\psi\rangle = (a_1|\uparrow\rangle + b_1|\downarrow\rangle)(a_2|\uparrow\rangle + b_2|\downarrow\rangle), \tag{2.4}$$

which is a product of a pair of two-dimensional Hilbert spaces, often written as $H_2 \otimes H_2$ (a composite product Hilbert space of a pair of two-dimensional Hilbert spaces, one for each qubit). But, this wave function (2.4) is *not* entangled. It can be separated into its two separate atomic parts. It is merely a superposition of two non-entangled atomic wave functions. The wave function (2.1) *cannot be separated* into its two component parts, and that is a signal for entanglement.

Many-Body Wave Functions. It was commented above, and in Chap. 1, that the interaction of the two particles created a many-body (in the discussion it was a two-body) wave function. Creating such a wave function is a common problem in quantum physics. With fermions (particles with half-integer spin), it is important that the wave function preserve the anti-symmetry when two particles are interchanged. One form of creating a many-body wave function is through the Slater determinant [6]. To illustrate this, consider a N-dimension Hilbert space to be occupied by N particles (fermions). To give each particle its own quantum state, we need a set of parameters whose combinations can yield N-possible combinations, which will be donated as the set of paramters $\{a_i\}$ with $i = 1, 2, ..., N$. The Slater determinant is then defined to be

$$|\psi_N\rangle = \frac{1}{\sqrt{N}} \begin{vmatrix} |a_1\rangle_1 & |a_2\rangle_1 & \cdots & |a_N\rangle_1 \\ |a_1\rangle_2 & |a_2\rangle_2 & \cdots & |a_N\rangle_2 \\ \vdots & \vdots & & \vdots \\ |a_1\rangle_N & |a_2\rangle_N & \cdots & |a_N\rangle_N \end{vmatrix}. \tag{2.5}$$

So, each row runs through the set of parameters for a given particle, and the columns run through the particles for a given parameter. The numerical factor in front of the determinant assures that the wave function magnitude is normalized to unity.

For the Bohm-Aharonov system, the Hilbert space has only two particles and two possible states of a single parameter. Then, the Slater determinant becomes

$$|\psi_2\rangle = \frac{1}{\sqrt{2}} \begin{vmatrix} |\uparrow\rangle_1 & |\downarrow\rangle_1 \\ |\uparrow\rangle_2 & |\downarrow\rangle_2 \end{vmatrix}. \tag{2.6}$$

If this determinant is expanded, it yields exactly (2.1). Thus, the proposed Bohm-Aharonov entangled state is precisely a Slater determinant which preserves the anti-symmetry of the fermions of which it is composed. Interchanging the two particles in (2.1) (that is, interchanging the subscripts) merely reverses the order of the two terms, and yields a minus sign in front of the right-hand side of the equation. The minus sign is the anti-symmetric result of the interchange of the two particles.

2.2   Two-Level Systems

In Chap. 1, the normal classical bit was described in terms of two levels, corresponding to the $|0\rangle$ and $|1\rangle$ states. Both there, and above, the qubit was also described in terms of two levels representing the same states, but as quantum states. These two levels can exist in all types of systems, as were described for the various types of qubits in Sect. 1.3. But, there are particular properties that are desired for use in qubits. For example, in a potential well imposed by surface gates on a semiconductor, the self-consistent potential that results tends to be of quadratic nature. This is unfortunate, as quadratic potentials in quantum mechanics give a set of energy levels that tend to be equally spaced (e.g., the harmonic oscillator, or pendulum classically). This is not good for qubit use, as choosing any two levels cannot avoid particles drifting off to other energy states, which constitutes a loss of information. On the other hand, the $1/r$ potential in an atom leads to an unequal spacing of the energy levels. Usually, the lowest levels are spaced the largest amount, and often are thus used for qubits. Hence, qubits are often referred to as atomic states, even when they are not, because of this preferred spacing in the energy levels.

Using these two states can lead to proper behavior in qubits; as a result of the above arguments, two-level systems have come to have their own well-developed properties. In this section, these properties, especially those suited for use with qubits, are further discussed in order to provide some indication of how qubits are used in processing. Consider Fig. 2.1, in which a two level system is described. In panel (a), the two levels are labeled as E_0 and E_1, in which the former is assumed to be the ground state and the latter is the next higher excited state. Obviously, these two states are separated by an energy difference $\Delta = E_1 - E_0$. This system of two levels can be probed by electromagnetic radiation. Whether this is radio frequency, microwave frequency, or infrared or optical frequency radiation all depends upon the size of Δ. The corresponding frequency is given by the Planck relation

$$f = \frac{\Delta}{h},\tag{2.7}$$

where h is Planck's constant ($\sim 6.626... \times 10^{-34} Joule - sec$; We will also use $\omega = 2\pi f$ and $\hbar = h/2\pi$). When the two-level system is irradiated by a signal with the proper frequency, an electron in the ground state can be excited to the upper energy state. This is shown by the green electron (dot) and arrow in panel (a). But, if the excitation signal remains on the system, the electron in the upper level can be induced to relax to the lower level (panel b). This is termed *stimulated emission*, a term coined by Einstein [7]. When stimulated emission occurs, then the incident photon is joined by a new photon emitted by the electron as it relaxes. This process provides the gain necessary for lasers [8]. Even if the signal is turned off, the electron in the upper state can relax to the ground state, since this is the normal equilibrium position. This relaxation is termed *spontaneous emission*, and it would be useful if this process didn't exist, at least for qubits where it is a loss

(a)   (b)

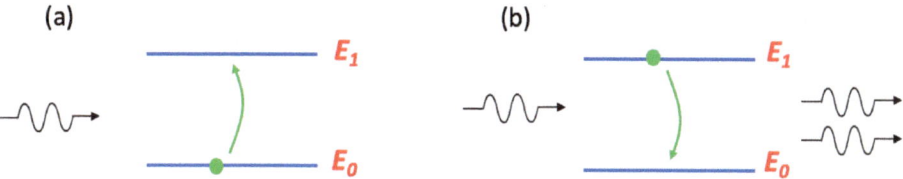

Fig. 2.1 Various operations can be performed on a two-level system. In **a**, a photon is being absorbed which excites an electron from the ground state to the excited state. In **b**, the photon induces the electron to decay to the ground state, thereby emitting an additional photon

mechanism. If the probability that the electron stays in the upper level (in the absence of the radiation) varies as $exp(-t/\tau)$, then τ is the lifetime, or inverse of the decay rate. So, for qubits, it is desired that the upper state has a very long lifetime, which is often where the upper state is described as metastable.

The above processes of absorption and emission lead us to see that there is a possibility that the electron will oscillate between the two levels if the signal remains applied to the system. The frequency of this oscillating electron is termed the Rabi frequency [9]. It is this steady oscillation that really opened the door to the use of two-level systems for qubits, because of the importance of the phase. To understand this, we have to dig a little deeper into the understanding of the two-level system.

Polarization. The two levels shown in Fig. 2.1 could as easily be thought of as the two spin states discussed in Sect. 2.1. To do this, it is easy to write the Hamiltonian (or energy operator, which describes the motion of the electron in the figure) as

$$H = \frac{\Delta}{2}\sigma_z, \tag{2.8}$$

where Δ is the separation between the two levels introduced above, and σ_z is one of the Pauli spinor matrices, specifically given as

$$\sigma_z = \begin{bmatrix} 1 & 0 \\ 0 & -1 \end{bmatrix}. \tag{2.9}$$

What has been done here is to move the zero-energy point to the mid-point between the two energy levels. Then, the upper level has the energy $\Delta/2$ and the lower level has the energy $-\Delta/2$. The presence of the matrix indicates that (2.8) refers to a two-level system in which spin notation is going to be used. That is, the states (0, 1) have been mapped into the states $(-1, 1)$ in terms of $\Delta/2$. In this new notation, the wave function (2.2) is now written as

$$\psi = \begin{bmatrix} a \\ b \end{bmatrix} = \begin{bmatrix} |a|e^{i\phi_1} \\ |b|e^{i\phi_2} \end{bmatrix}, \tag{2.10}$$

so that each of the two coefficients has its own phase.

It is useful to write the two wave functions as a single density matrix ($\rho = \psi\psi^\dagger$, where $\dagger$ indicates the adjoint, which would be the complex conjugate in one dimension)

$$\rho = \begin{bmatrix} a \\ b \end{bmatrix} \begin{bmatrix} a^* & b^* \end{bmatrix} = \begin{bmatrix} |a|^2 & ab^* \\ a^*b & |b|^2 \end{bmatrix}. \tag{2.11}$$

(Here, it is clear that the adjoint operation, when used on matrices, implies using the transpose of the matrix plus the complex conjugates of the elements.) The trace of such a matrix is the sum of the diagonal components, so that the trace of (2.11) satisfies (2.3) as

$$Tr\left\{ \begin{bmatrix} |a|^2 & ab^* \\ a^*b & |b|^2 \end{bmatrix} \right\} = |a|^2 + |b|^2 = 1. \tag{2.12}$$

So, this approach continues to assure that the magnitude of the wave function is unity and all is consistent.

It has been established over the years that this density matrix has an interesting property that can be exploited here to begin to understand how qubits operate. Unfortunately, this will introduce the full set of Dirac spinors, described below. This important property is that (2.11) can be rewritten in a different form by introducing the *polarization P* (also called the Bloch vector), such that

$$\rho = \frac{1}{2}(I + \boldsymbol{P} \cdot \boldsymbol{\sigma}), \tag{2.13}$$

where I is the 2×2 identity matrix, and $\boldsymbol{\sigma}$ is the Pauli vector

$$\boldsymbol{\sigma} = \begin{bmatrix} 0 & 1 \\ 1 & 0 \end{bmatrix} \boldsymbol{a}_x + \begin{bmatrix} 0 & -i \\ i & 0 \end{bmatrix} \boldsymbol{a}_y + \begin{bmatrix} 1 & 0 \\ 0 & -1 \end{bmatrix} \boldsymbol{a}_z, \tag{2.14}$$

and the $\boldsymbol{a}_i$ are unit vectors along the three coordinate axes. The three matrices are the three Pauli spinors for the three directions. By comparing (2.13) to (2.12), it is possible to identify the three parts of the polarization vector as

$$\begin{aligned} P_x &= 2Re\{a^*b\} \\ P_y &= 2Im\{a^*b\} \\ P_z &= |a|^2 - |b|^2 \end{aligned} \tag{2.15}$$

The language that has been developed above allows one to express the polarization vector in a real three-dimensional cartesian coordinate system. But, it can also be expressed as a vector in an abstract Euclidean space known as the Bloch sphere, shown in Fig. 2.2. Here, the polarization has unit amplitude and is shown in a spherical coordinate system. The vector can be projected onto the cartesian coordinates using the angles θ, ϕ shown in the figure. Many operations can be performed on $\boldsymbol{P}$, while retaining its amplitude on

Fig. 2.2 The Bloch sphere represents the polarization of the two-level system in a spherical coordinate system. The projection of the polarization vector onto the cartesian axes is determined by the two angles defined in the figure. The sphere itself has unit radius to preserve the normalization of the wave function

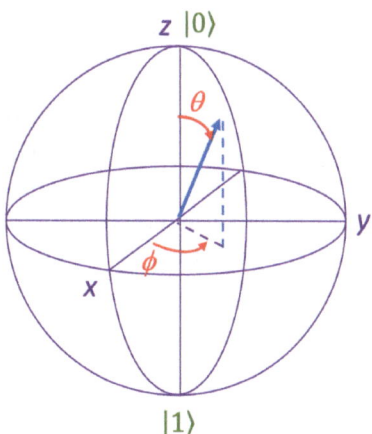

the surface of the unit sphere. Natural rotations around any of the three cartesian axes are common operations performed on qubits. There are others besides these rotations. Similar to (2.15), the three components of the polarization can be written in terms of the two angles as

$$P_x = P\sin(\theta)\cos(\phi)$$
$$P_y = P\sin(\theta)\sin(\phi) \tag{2.16}$$
$$P_z = P\cos(\theta).$$

Notice that the two qubit states are defined along the z-axis, with the $|0\rangle$ in the positive z-direction. Thus, switching from the $|0\rangle$ state to the $|1\rangle$ state involves a π rotation about either the x or the y axis. That is, quite generally, rotations away from the positive or negative z-axis are creating superposition states of the two basis states of the qubit. When talking about *preparing* the qubit, it means that initial rotations are performed to obtain a desired superposition at the start of a computation.

Manipulating P. It should now be clear that performing operations on a qubit is really manipulating the polarization vector P. To understand how this is done requires a bit more mathematical development. To begin, it is necessary to understand that the diagonal terms of the density matrix (2.11) provide the probability, or square magnitude of the wave function, throughout the physical space. Thus, computing an average, or expectation, value of a parameter A is found by averaging it over this physical space, which is defined in terms of the density matrix as

$$\langle A \rangle = Tr\{\rho A\}. \tag{2.17}$$

As a result, the expectation value of one of the Pauli matrices is given as

$$\langle \sigma_x \rangle = Tr\{\rho\sigma_x\} = \frac{1}{2}(Tr\langle\sigma_x\rangle + \boldsymbol{P} \cdot Tr\{\sigma_x\boldsymbol{\sigma}\}), \tag{2.18}$$

where (2.13) has been used for the density matrix. At this point, various products of the Pauli matrices have to be determined. For example, one can write one such product as

$$\sigma_x\sigma_y = \begin{bmatrix} 0 & 1 \\ 1 & 0 \end{bmatrix}\begin{bmatrix} 0 & -i \\ i & 0 \end{bmatrix} = \begin{bmatrix} i & 0 \\ 0 & -i \end{bmatrix} = i\sigma_z. \tag{2.19}$$

Other products can be obtained by permuting the various indices (in a right-hand sense). Note that the trace of these cross-products is zero. On the other hand,

$$\sigma_x\sigma_x = \begin{bmatrix} 0 & 1 \\ 1 & 0 \end{bmatrix}\begin{bmatrix} 0 & 1 \\ 1 & 0 \end{bmatrix} = \begin{bmatrix} 1 & 0 \\ 0 & 1 \end{bmatrix} = I. \tag{2.20}$$

This means that (2.18) becomes

$$\langle\sigma_x\rangle = Tr\{\rho\sigma_x\} = P_x. \tag{2.21}$$

This is an interesting result, since it establishes that the polarization vector actually indicates the direction of the expectation value of the spin vector in space. Another important point lies in (2.19), where if the two matrices are reversed, it yields a minus sign in the result–the order of the two matrices is important; they are said to not commute (that is, two matrices A and B that do not commute satisfy $AB \neq BA$).

Now suppose that a force is applied to the two-level system. It may be assumed that there is a scalar force Q_0 and a vector force Q. In this situation, and these forces are contained in a Hamiltonian given by

$$H = \frac{1}{2}(Q_0I + Q \cdot \sigma). \tag{2.22}$$

The reason for this form will become apparent in the discussion below. In quantum mechanics, the rate of change of a parameter is given by its commutator with the Hamiltonian acting upon it, or

$$\frac{dP}{dt} = \frac{d\langle\sigma\rangle}{dt} = \frac{1}{i\hbar}[\sigma H - H\sigma]. \tag{2.23}$$

When the Hamiltonian of (2.22) is used, the first term (the identity matrix) commutes and vanishes from the equation. Only the second term remains. This now leads to

$$\frac{d\langle\sigma\rangle}{dt} = \frac{1}{i2\hbar}\langle\sigma(Q \cdot \sigma) - (Q \cdot \sigma)\sigma\rangle = \frac{1}{i2\hbar}\langle Q \times (\sigma \times \sigma)\rangle. \tag{2.24}$$

In normal situations, the last cross product on the right-hand side ($\sigma \times \sigma$) of this equation would vanish, since normal scalars commute. But, these Pauli matrices do not commute, as demonstrated in (2.19). Thus, one obtains

$$\boldsymbol{\sigma} \times \boldsymbol{\sigma} = \left(\sigma_x \boldsymbol{a}_x + \sigma_y \boldsymbol{a}_y + \sigma_z \boldsymbol{a}_z\right) \times \left(\sigma_x \boldsymbol{a}_x + \sigma_y \boldsymbol{a}_y + \sigma_z \boldsymbol{a}_z\right)$$
$$= \left(\sigma_y \sigma_z - \sigma_z \sigma_y\right)\boldsymbol{a}_x + \left(\sigma_z \sigma_x - \sigma_x \sigma_z\right)\boldsymbol{a}_y + \left(\sigma_x \sigma_y - \sigma_y \sigma_x\right)\boldsymbol{a}_z \qquad (2.25)$$
$$= 2i\boldsymbol{\sigma}$$

where the products from (2.19) have been used. Thus, the cross product of the spin vector with itself yields only a factor of $2i$ times the spin vector. This is another peculiar behavior of spin. This now takes (2.24) into the form

$$\frac{d\boldsymbol{P}}{dt} = \frac{d\langle\boldsymbol{\sigma}\rangle}{dt} = \frac{1}{\hbar}\boldsymbol{Q} \times \langle\boldsymbol{\sigma}\rangle. \qquad (2.26)$$

What this tells us is that the polarization vector precesses around the force vector $\boldsymbol{Q}$. If there is a need to rotate P around the x-axis, a force along this axis is used. This holds for the other axes as well. The radian frequency of this precession is given by

$$\omega_Q = \frac{1}{\hbar}\boldsymbol{Q}. \qquad (2.27)$$

Note that there is no rotation if the polarization vector is parallel to the force vector. As a result, for a given force vector, there are two stable positions in which there is no precession, and the polarization in unaffected. These are

$$\boldsymbol{P} = \pm\boldsymbol{a}_Q, \qquad (2.28)$$

in which $\boldsymbol{a}_Q$ is a unit vector in the direction of the force. Each of these two states has an energy (related to the square magnitude of the polarization) given as

$$\frac{1}{2}(Q_0 \pm Q), \qquad (2.29)$$

although only the difference

$$\Delta E = Q = \hbar\omega_Q \qquad (2.30)$$

is important. If the force is a magnetic field, this splitting gives the Zeeman splitting that is expected from a magnetic field. These forces do not change the amplitude of the polarization, only its direction. So, these forces are just those needed to manipulate the qubit.

Qubits. At this point, it is perhaps appropriate to get a better feeling on the two angles that govern the direction of the polarization vector and the rotations that might be involved. Consider, for example, the state

$$\frac{1}{2}\left(\sqrt{3}|0\rangle + |1\rangle\right), \qquad (2.31)$$

with the prefactor allowing the proper normalization. Now, projection onto the z-axis gives the first value so that, from (2.16), one has

$$cos(\theta) = \frac{\sqrt{3}}{2}, \theta = \frac{\pi}{6}. \tag{2.32}$$

Note that, with the current notation, this latter angle is constrained to lie in the range $0 \leq \theta \leq \pi$, as is normal in spherical coordinates [10] (many in the quantum computing community treat the value here as $\theta/2$ rather than θ, but that is inconsistent with normal spherical coordinates). The 1 state fraction is given as

$$e^{i\phi} sin(\theta) = \frac{1}{2}, \tag{2.33}$$

and with $sin(\pi/6) = 1/2$, we are left with $e^{i\phi} = 1$, or $\phi = 0$. This vector lies in the (x,z) plane at an angle of 30° from the z-axis toward the $+x$ direction. As a second example, consider the state

$$\frac{1}{\sqrt{2}}(|0\rangle - |1\rangle), \tag{2.34}$$

So that

$$cos(\theta) = \frac{1}{\sqrt{2}}, \theta = \frac{\pi}{4}. \tag{2.35}$$

Now, if we consider

$$e^{i\phi} sin(\theta) = -\frac{1}{\sqrt{2}}, \tag{2.36}$$

then with $sin(\pi/4) = 1/\sqrt{2}$, one obtains $e^{i\phi} = -1$, or $\phi = \pi$. This vector also lies in the (x, z) plane at an angle of 45° from the z direction toward the $-x$ direction.

It is apparent from the previous section that operations on the qubit mean changing its state and, in particular, its phases. There are (at least) two ways of doing this in what might be called a normal operating state. These can be described as (1) continuous excitation, and (2) pulsed excitation. Using these terms may be more fanciful than the world is with quantum information, but they cover two of the basic methods one can think of with which to change the phase of the two-level atom.

With continuous excitation, the radiation signal shown as the photon on the left of Fig. 2.1a is applied to the atom continuously so that there is a continuous Rabi oscillation of the electron between the two levels. In this approach, the phase of the Rabi oscillation, with respect to a reference phase, perhaps provided by the excitation radiation, gives the important state of the qubit. This is going to require careful control of the phase of the excitation and of the phase of the qubit.

In pulsed excitation, as may be assumed, the photon source provided by the exciting radiation is not continuous, but is provided under pulsed condition. That is, the radiation is provided for a well-controlled time duration, which is termed the pulse length. The transition between the two states occurs at the Rabi frequency. This frequency is related to the excitation frequency, but also to the separation of the two levels, the power provided by the excitation, the coupling between the excitation and the atom, and other material parameters (this will be discussed further below). If the Rabi radian frequency is taken to be $\Omega = 2\pi f_{Rabi}$, then an example of a π-pulse would have a period given as

$$T = \frac{\pi}{\Omega} = \frac{1}{2f_{Rabi}}. \tag{2.37}$$

Such a pulse would just excite the electron from the ground state to the excited state, or vice versa, corresponding the π rotation discussed above. This π rotation of the polarization vector, for example transforms from the $|0\rangle$ to the $|1\rangle$ state on the Bloch sphere of Fig. 2.2 (or vice versa).

When power is mentioned, this mainly refers to the rate of photon arrival; e.g., how many photons per second impinge upon the atom. Whether it is microwaves or optical waves, the radiation signal is characterized by energy and power. The photons have an energy given by Planck's relationship $E = hf = \hbar\omega$, and the signal has a power given by Joules/s or Watts. For example, station WSM, an AM station in Nashville, TN, delivers 50,000 watts at a frequency of 650 kHz. This means that each photon has about $4.310^{-28} Joules$ or $2.7 \times 10^{-9} eV$. The power level means that it delivers about 1.2×10^{33} photons/s. On the other hand, a blue laser operates at a wavelength around 400 nm, which means that the photons have an energy near 3.1 eV. If the laser delivers 5 mW, then the photon flux is about 10^{16} photons/s, which is still a substantial rate. Under each of these two conditions, it would be difficult to distinguish individual photons. In the latter case, a photon arrives roughly every 0.1 fs. This would be an exceptionally short pulse.

Synchronization. In the above discussion, it was clear that one needs to maintain close control over frequencies, phases, pulse lengths, and so on. This is no different in quantum information than in classical information. In classical computing, there is a system clock which governs overall processing rates. There may be additional local clocks which govern certain internal chip processes. Nevertheless, all of these clock signals must be carefully controlled and synchronized. If one wants to read a data value (the data word which may be 8, 16, 32, 64 or more bits), the control unit must take the address assigned by the program and convert this to the particular spatial position of those bits in the memory system, and then send this address to the memory controller, and this is done at a particular clock pulse. The memory controller must then enable those sites to be read on a later clock pulse. The memory controller likely puts these bits into a memory register which then shifts the bits onto the data bus as the word is transmitted serially to the main processor. Even simple arithmetic needs this synchronization. If one wants to add two numbers A and B, A must be read from memory (or perhaps from a temporary *cache* memory on chip) and placed into a register in the central processing unit. *B* is similarly placed in another register. If the register where A is located is capable of simple arithmetic, then *B* can be added to it with the result staying in this register. If not, everything is transferred to the arithmetic logic unit. This all takes several clock cycles, yet it must all be synchronized. Even a simple gate has this problem, as the signals to the gate are prepared (perhaps in a storage register), and then the gate operation is enabled by a subsequent clock signal. This enable signal must be present sufficiently long for the transistors to switch in the gate. *Synchronization is the single most important detail of computing!* On modern microprocessors which have relatively large chip sizes, delivering the clock precisely on time to each and every part of the chip is a modern technology problem (some refer to it as a nightmare). Clock skew (as slips in the clock at different points on a chip is called) can kill the performance of a chip. When one says that the devil is in the details, synchronization is that devil!

In quantum information, synchronization is no less important. Here, it is referred to as quantum control. The precise preparation of initial qubit states, the actual operation of qubit gates, the timing of entangled gates/particles, must all be determined and synchronized by quantum control processes. Because quantum processing does not have the natural restoration of logic values that occur with transistors, quantum control is perhaps much more difficult and necessary. This suggests that quantum control difficulties are likely to be the same classical control problems on steroids!

2.3   Addressing the Qubit–The Jaynes-Cummings Model

In Chap. 1, there was considerable discussion of using electromagnetic waves to couple to the atomic levels of possible qubits, whether these qubits were real atoms or atom-like Josephson junction qubits, or any other type of qubit. In nearly all cases, radiation is used to manipulate the qubit in one form or another. Here, we want to discuss the actual physics of coupling the radiation to the atomic-like levels of the qubit. The coupling between a two-level "atom" (our two-level qubit of the previous section) and a resonant (quantized) electromagnetic cavity is described by a relatively old model, termed the Jaynes-Cummings model [11]. The process of coupling coherent radiation to such an atom is of course important in quantum optics, but it is a general model that can be applied to nearly all qubits, especially when they are coupled to radiation.

The model is depicted in Fig. 2.3, where the atom is considered as a two-level system. This atom interacts with a quantized mode of an electromagnetic cavity. The model is actually very analogous to two coupled classical pendula, in which the interaction leads to an interaction between the two oscillator modes in a way in which all of the energy itself oscillates between the two pendula. At a given instant of time, pendulum 1 may be static while all the energy is in the oscillation of pendulum 2, and at a later time the process is reversed with all the energy in pendulum 1. In the quantum case, one pendulum is

Fig. 2.3 A schematic diagram of the Jaynes-Cummings model, in which an electromagnetic cavity is coupled to a two-level atom (the qubit)

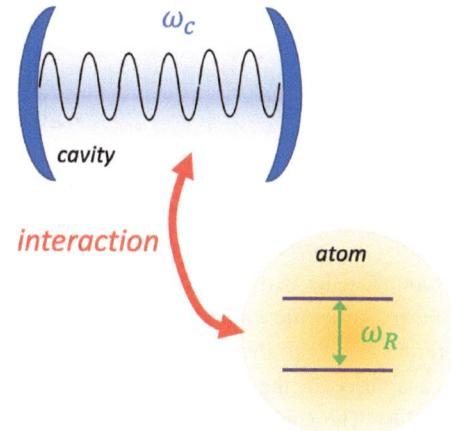

the harmonic oscillator mode of the electromagnetic cavity, while the second is the atom, which undergoes its Rabi oscillations. As with the pendulum, the Rabi oscillation is not static in amplitude, but varies as the energy moves back and forth between the cavity and the atom.

Just as in the coupled pendula, as the Rabi oscillations will collapse and then reappear periodically, the energy moves back and forth between the cavity and the atom. The temporal time over which this collapse and reappearance occurs has been termed the revival time [12]. If the cavity is in a coherent state, and the atom is prepared in an excited state, the atom and field states become rapidly entangled, and then subsequently disentangle at one-half the revival time [13]. That is, the maximum entanglement of the atomic states occurs when the Rabi oscillations are strongest and the minimum entanglement occurs when the Rabi oscillations have vanished. Recently, the photon parity operator has been shown to also have similar behavior, but with one important difference. The peaks of the amplitude of the parity operator seem to occur when the Rabi oscillations have vanished. The amplitude and decay of this parity operator seems to be shifted by half a period from the Rabi oscillations [14]. The parity operator is proportional to the density of photons in the radiation field, which also correlates with the two-pendulum system. This is because the Rabi oscillation vanishes when the totality of the energy is in the photon oscillator. The Jaynes-Cummings model was revived in this century [15] and applied explicitly to the problem of a trapped ion interacting with a laser field (which was mentioned in Chap. 1 in connection with one of the commercial approaches to a quantum computer).

In order to study this model, one needs the Hamiltonian of the system. This can be constructed in terms of the various components shown in Fig. 2.3. First, the electromagnetic field is quantized as usual in terms of a series of harmonic oscillators, one oscillator for each of the modes that can exist in the cavity [16]. In the case of interest, only a single mode is considered here. For this single mode, the wave energy may be written as[1] [17, 18]

$$H_{EM} = \hbar\omega_C \hat{a}^\dagger \hat{a}, \tag{2.38}$$

where the operators are the normal creation and annihilation operators for the field in the harmonic oscillator representation of the cavity, and ω_C is the frequency of this mode (that is, the operator $\hat{a}^\dagger$ creates a photon in this mode, while $\hat{a}$ destroys a photon, and the product of the two $\hat{a}^\dagger \hat{a}$ is just the number of photons in the mode; this product is often referred to as the *number operator*, but it is not an operator at all, just a number). The atomic levels are the two levels of the atom, and are described with a pseudo-spin index just as used in the polarization vector of the qubit on the surface of the Bloch sphere in (2.8). The spin-up state refers to the upper level and the spin-down state refers to the lower level. The atomic Hamiltonian then may be written as

$$H_a = \hbar\omega_R \frac{\sigma_z}{2}, \tag{2.39}$$

just as in (2.8), where here $\Delta \rightarrow \hbar\omega_R$. Hence, the atomic frequency ω_R describes the separation of the two levels. The state of the atom can then be described simply by its polarization

$$\hat{P} = \hat{b}^\dagger + \hat{b}, \tag{2.40}$$

where $\hat{b}^\dagger$ and $\hat{b}$ are the normal raising and lowering operators of the two-level atom in its own harmonic oscillator notation (here $\hat{b}^\dagger + \hat{b}$ are used rather than $\hat{a}^\dagger$ and $\hat{a}$, because the latter two have already been used for the cavity photons). Hence, the polarization here is just the position x in a normal harmonic oscillator. The various parameters will be folded into a pre-factor as discussed below.

The electromagnetic field and the atom interact through the electric field of the electromagnetic wave. The potential that arises from the field E is eEx, so here x is replaced by the polarization. Then the interaction energy is

$$H_{int} = \frac{\hbar\Omega}{2} E_x \hat{x} \hat{P} = \frac{\hbar\Omega}{2} E_z \left(\hat{a}^\dagger + \hat{a}\right)\left(\hat{b}^\dagger + \hat{b}\right). \tag{2.41}$$

The constant Ω incorporates all of the various constants that arise from the connection between the operators that appear in the previous equations. The creation and annihilation operators each have their time variation determined by their respective frequencies as in the normal case. The four cross terms in (2.41) have slow variations due to the difference

---

[1] The harmonic oscillator (see Appendix A for a discussion of this system and explanation of the operators used) is one of the few completely solvable problems in quantum mechanics, and its solution using the operators of (2.38), which greatly simplify the mathematics is usual and can be found in most introductory textbooks, some of which are cited above.

in the frequencies and fast variations due to the sum of the two frequencies. Generally, one adopts the rotating-wave approximation in which the fast components are ignored. (The equivalent step in the coupled pendula is to ignore the actual oscillations of the two pendula and concentrate on the slower movement of energy back and forth between the two.) The Hamiltonian is then separated into two commuting parts, and these are described by the two equations

$$
\begin{aligned}
H_1 &= \hbar \omega_C \left(\hat{a}^\dagger \hat{a} + \tfrac{\sigma_z}{2} \right), \\
H_2 &= \hbar \delta \tfrac{\sigma_z}{2} + \tfrac{\hbar \Omega}{2} \left(\hat{a} \hat{b}^\dagger + \hat{a}^\dagger \hat{b} \right).
\end{aligned}
\tag{2.42}
$$

In this equation, the term $\delta = \omega_R - \omega_C$ is the detuning of the frequency between the atom and the field. The first of these equations is just the energy in the radiation-like oscillations at the two atomic levels. The second equation actually deals with the coupling of the atom to the cavity oscillation. These two equations can be solved together and this will yield two hybrid energies for the interacting system (hybrid because they are combinations of the atom and field frequencies, and are different from either of these two separate frequencies). The operator product $\hat{a}^\dagger \hat{a}$ is the number of photons n, as mentioned previously, so that these hybrid frequencies are given as

$$
E_\pm = \hbar \omega_C \left(n + \frac{1}{2} \right) \pm \frac{\hbar \Omega_n(\delta)}{2},
\tag{2.43}
$$

where the effective Rabi frequency is now

$$
\Omega_n = \sqrt{\delta^2 + \Omega^2 n + 1}.
\tag{2.44}
$$

When the detuning is small, the atomic state Rabi oscillations occur with an approximate frequency $\Omega_C = 2\Omega\sqrt{n+1}$. As the electromagnetic mode occupation builds up (the density n increases), the Rabi frequency also increases (the amplitude of these oscillations increases as the pumping power from the radiation field is increasing, thus driving the atomic transition to happen quicker). The photon parity operator is a useful quantity that is defined as

$$
\Pi_F = (-1)^n = e^{in\pi},
\tag{2.45}
$$

where the n here is the number of photons, just as above. This parity operator will appear later in some measurements schemes described below.

When the wave functions are taken into account, the expectation value (its quantum mechanical average) of the parity operator may be shown to be [14]

$$
\langle \Pi_F(t) \rangle = e^{-n} \sum_{s=0}^{\infty} (-1)^s \frac{n^s}{s!} \cos \left(2\Omega t \sqrt{s+1} \right),
\tag{2.46}
$$

in which the summation is over the possible configurations of the system. Now, at $t = 0$, the cosine term is unity, and the parity is dominated by the exponential term, which goes to zero as the photon number increases. Thus, when Rabi oscillation is maximum, the parity has vanished. Indeed, the parity is maximum when the Rabi oscillation is minimum (n goes to zero).

A great many variations in the Jaynes-Cummings model have been considered. A generalization of the model through the use of a Markovian master equation approach, that replaces the Hamiltonian form, has been developed [19, 20]. The model also has been extended to a double Jaynes-Cummings model, that incorporates a pair of two-level atoms [21–23], and even to a pair of three-level atoms [24]. An anti-Jaynes-Cummings model has appeared in which the two field operators in the second term of H_2 in (2.42) are interchanged [25] (remember that these operators do not commute). These various forms have allowed the use of the Jaynes-Cummings model in a variety of qubit formulations.

2.4   Qubit Gates

Quantum computers work with qubit gates that are controlled as needed, such as through the Jaynes-Cummings model. This in fact is little different from classical computers. In both cases, these gates provide the interaction between the qubits or bits, both of which denote a logic value. Here, the CNOT gate will be discussed as an example (CNOT is short for Controlled NOT). Such a gate is shown in Fig. 2.4a schematically. The "signal" to be inverted is labeled as the input x (in green). The control signal is labeled c (in red). The principle is that if the control input is a 1, the signal is inverted, either changed from 0 to 1 or from 1 to 0. If the control input is a 0, nothing happens and the signal is unchanged. This is depicted in the so-called truth table of panel (b) of the figure. Perhaps this is called a truth table because of Boole's original use of the binary system as an indicator of truth in philosophical arguments [26]. Here, a 1 indicates a true signal, while 0 indicates a false signal.

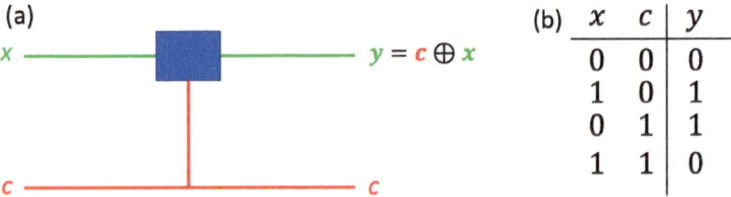

Fig. 2.4 **a** The CNOT gate is depicted schematically. **b** The truth table for the gate. The inputs are in the first two columns (as labeled), while the output is in the last column. This gate has many other names in classical computing, as discussed in the text

However, the CNOT gate is not unique to quantum computing. It appears in many forms in classical computers. Classically, any transistor (or equally any complementary metal–oxide–semiconductor, termed CMOS, gate) automatically is an inverting gate, or NOT gate. Since the transistor inverts its input, a high voltage input (which is taken as a logical 1) turns on the transistor, and causes the output voltage to become low (which is taken to be a logical 0). Similarly, when the input is low, the output is high. So, this transistor is a natural NOT gate. The CNOT version is slightly different and requires more transistors (or gates), since the logic is that the output is 1 whenever either of the two inputs is 1, but not when both are 1. Classically, this is called an exclusive OR (denoted as XOR) circuit. The truth table is the same as Fig. 2.4b. But, this can also be the arithmetic part of what is called a *half*-adder. In binary arithmetic, if one ignores the "carry" operations, then $1 + 1 = 0$. (This is also true in normal base-10 arithmetic where $5 + 5 = 0$.) But life seems to need us to keep track of those carries. Thus, the *full* adder for a pair of bits allows for a carry-in from the next lower-order bit and for a carry-out to the next higher-order bit, just as taught in the first experience of arithmetic. In the classical case, the control is the clock pulse that enables the addition to proceed.

Returning to the quantum version of the gate, one can see that the output y is unity when the input x is 0 and the control is 1, or when the input x is 1 and the control is 0. If the input state is taken as the first "particle" and the control states as the second "particle," then the output wave function can be written as

$$y \rightarrow \psi = \frac{1}{2}(\lceil 0 \rangle \lceil 1 \rangle \pm \lceil 1 \rangle \lceil 0 \rangle), \tag{2.47}$$

which is just (2.1) with a different notation. Hence, it is clear that this quantum gate entangles the two inputs. It is this property that distinguishes the quantum gate from a classical one.

Gate Operations. As noted, the CNOT quantum gate is operating on a pair of qubits (just as in the classical case); both the signal and the control inputs come from the two qubits holding these signals. There are a variety of gate operations, some working on a single qubit and some working on a pair of qubits. One can envision various operations which work on a larger number of qubits, but in most cases they can be expressed as a combination of single qubit operations and two qubit operations. In Chap. 1, the idea of a state transition matrix was introduced, and this matrix notation can be used here. To better understand this, let us return to the notation (2.10). If the system is in the 0 state, then $a = 1$ and $b = 0$, so that

$$|0\rangle = \begin{bmatrix} 1 \\ 0 \end{bmatrix}, \tag{2.48}$$

in which the wave function is now represented as a "spinor" state (a general name for a two-element matrix representation, derived from Pauli's spin descriptions). Similarly, when the system is in the 1 state, then $a = 0$ and $b = 1$, so that

$$|1\rangle = \begin{bmatrix} 0 \\ 1 \end{bmatrix}. \tag{2.49}$$

As mentioned, there are a variety of single qubit gates, some of which will be illustrated here. The first is a unity gate, which actually does nothing. It may be represented by the operation (I is the two-level unit matrix, denoted as the last matrix)

$$O = I = \sigma_0 = \begin{bmatrix} 1 & 0 \\ 0 & 1 \end{bmatrix}, \tag{2.50}$$

so that

$$O\psi = \begin{bmatrix} 1 & 0 \\ 0 & 1 \end{bmatrix} \begin{bmatrix} a \\ b \end{bmatrix} = \begin{bmatrix} a \\ b \end{bmatrix}. \tag{2.51}$$

While this gate does nothing, it illustrates how the qubit operations are to be interpreted within this mathematical approach, and it will actually be used in some encryption methods, as discussed below. Now, consider an X rotation which is naturally defined by the Pauli σ_x, so that

$$X\psi = \begin{bmatrix} 0 & 1 \\ 1 & 0 \end{bmatrix} \begin{bmatrix} a \\ b \end{bmatrix} = \begin{bmatrix} b \\ a \end{bmatrix}. \tag{2.52}$$

In essence, this is the same result as the NOT gate, in that the two values are inverted. On the other hand, a Y operation proceeds as

$$Y\psi = \begin{bmatrix} 0 & -i \\ i & 0 \end{bmatrix} \begin{bmatrix} a \\ b \end{bmatrix} = i \begin{bmatrix} -b \\ a \end{bmatrix}. \tag{2.53}$$

This is a much more complicated operation that changes both of the angles defined on the Bloch sphere as the polarization is rotated around the y-axis. The Z operation is again relatively simple involving just a rotation of ϕ by a factor of π. But, these are not the normal rotations.

Consider the action of the iY gate on the qubit (2.31), which may be rewritten here in the notation introduced above as

$$\frac{1}{2}\left(\sqrt{3}|0\rangle + |1\rangle\right) = \frac{\sqrt{3}}{2} \begin{bmatrix} 1 \\ 0 \end{bmatrix} + \frac{1}{2} \begin{bmatrix} 0 \\ 1 \end{bmatrix}. \tag{2.54}$$

Operating with the iY gate on this yields the result

$$-\frac{\sqrt{3}}{2} \begin{bmatrix} 0 \\ 1 \end{bmatrix} + \frac{1}{2} \begin{bmatrix} 1 \\ 0 \end{bmatrix} = \frac{1}{2}\left(|0\rangle - \sqrt{3}|1\rangle\right). \tag{2.55}$$

For this result, one obtains $\theta = \pi/3$, $\phi = \pi$, which means that the polarization has been rotated around the y-axis by $\pi/2$, a reasonable result. But, this was not simply a rotation in the polar angle, but involved rotation in both angles. To obtain a simple rotation only of the polar angle, one needs to find the direction of the line A from the origin O out through the angle ϕ and then rotate around a line normal to the (O, A)-plane. On the other hand, a simple rotation of only ϕ involves a rotation around the z-axis. These are just peculiarities of the spherical coordinate system used for the Bloch sphere.

In general, a rotation about one of the axes is going to be defined somewhat differently (and in a more complicated manner). A general rotation around one of the axes may be defined as

$$R_i(\beta) = cos(\beta)I - isin(\beta)\sigma_i,\tag{2.56}$$

where $i = x, y, z$, as required. As an example, consider the y-axis rotation which becomes

$$R_y(\beta) = \begin{bmatrix} cos(\beta) & 0 \\ 0 & cos(\beta) \end{bmatrix} - i\begin{bmatrix} 0 & -isin(\beta) \\ isin(\beta) & 0 \end{bmatrix}.$$
$$= \begin{bmatrix} cos(\beta) & -sin(\beta) \\ sin(\beta) & cos(\beta) \end{bmatrix}.\tag{2.57}$$

Similarly, other rotations can be evaluated. (It may be noted that the sine terms have the opposite sign from typical coordinate rotations in classical mechanics [27], but this is because it is the wave function that is being rotated here, not the coordinate system.)

It is also common practice that product operations are allowed, such as $R_y(\beta)R_x(\gamma)$, where the two angles are usually different from each other. The inverse operations also exist and $R_i(\beta)R_i(-\beta) = I$. Some additional common qubit operations also exist, and find common usage. Among these are the Hadamard transform

$$H = \frac{1}{2}\begin{bmatrix} 1 & 1 \\ 1 & -1 \end{bmatrix},\tag{2.58}$$

and the "phase" transform

$$S = \begin{bmatrix} 1 & 0 \\ 0 & i \end{bmatrix}.\tag{2.59}$$

The Hadamard gate is useful for optical qubits, as it is essentially a 50% beam splitter, that is splitting an incident optical beam into two optical beams of equal amplitude.

With the CNOT gate, there are two qubits arriving at the input, and this leads to a different vector formed from the product of the two separate Hilbert spaces, which may be written as $H_2^a \otimes H_2^c$. To better understand the new vector, let us write the signal and control inputs as the spinors

$$x = \begin{bmatrix} a \\ b \end{bmatrix}, c = \begin{bmatrix} 0 \\ 1 \end{bmatrix}. \tag{2.60}$$

Then, the four element state vector may be written as

$$\begin{bmatrix} a \\ b \\ a \\ b \end{bmatrix} or \begin{bmatrix} 0a \\ 0b \\ 1a \\ 1b \end{bmatrix}, \tag{2.61}$$

where the top two values arise for when the control signal is 0 and the bottom two is for the control signal 1. The second form retains both qubit values, and is used below in some operations for clarity. Since the top two are unchanged in the CNOT, while the bottom two are interchanged, the state transition matrix for the CNOT may be expressed as, but we must recall that it operates on the second element of the right-hand side of (2.61),

$$\begin{bmatrix} 1 & 0 & 0 & 0 \\ 0 & 1 & 0 & 0 \\ 0 & 0 & 0 & 1 \\ 0 & 0 & 1 & 0 \end{bmatrix}. \tag{2.62}$$

When the qubit is simple (as used in some following examples of qubit transmission), the CNOT is clear and simple. What happens if it is applied to the qubit (2.31), which may be written as, in the current notation

$$\begin{bmatrix} 0\frac{\sqrt{3}}{2} \\ 0\frac{1}{2} \\ 1\frac{\sqrt{3}}{2} \\ 1\frac{1}{2} \end{bmatrix}. \tag{2.63}$$

Operating with the CNOT gate produces

$$\begin{bmatrix} 0\frac{\sqrt{3}}{2} \\ 0\frac{1}{2} \\ 1\frac{1}{2} \\ 1\frac{\sqrt{3}}{2} \end{bmatrix}, \tag{2.64}$$

So that the output of the CNOT signal line is

$$y = \frac{1}{2}\left(|0\rangle + \sqrt{3}|1\rangle\right), \tag{2.65}$$

for which $\theta = \pi/3$, $\phi = 0$. In this case, the CNOT produces only a $\pi/3$ rotation around the y-axis.

As a second example, consider the qubit (2.34). The input and control signals are now.

$$\begin{bmatrix} 0\frac{1}{\sqrt{2}} \\ 0\frac{-1}{\sqrt{2}} \\ 1\frac{1}{\sqrt{2}} \\ 1\frac{-1}{\sqrt{2}} \end{bmatrix}. \tag{2.66}$$

Operating with the CNOT now produces the output vector

$$\begin{bmatrix} 0\frac{1}{\sqrt{2}} \\ 0\frac{-1}{\sqrt{2}} \\ 1\frac{-1}{\sqrt{2}} \\ 1\frac{1}{\sqrt{2}} \end{bmatrix}. \tag{2.67}$$

Hence the output of signal line of the CNOT gate is the qubit

$$y = \frac{-1}{\sqrt{2}}(|0\rangle - |1\rangle). \tag{2.68}$$

This result lies on the surface of the Bloch sphere with $\theta = 3\pi/4$, $\phi = 0$, so that the polarization vector has been inverted through the origin, now pointing in the opposite direction.

In the case where the input is simply a 0 or a 1, the polarization vector is inverted. This simple inversion also happened for the second case of the qubit (2.34). But, for the qubit of (2.31), this was not the case. Here the polarization vector was simply rotated around the y-axis by $\pi/6$, leaving the value of ϕ unchanged. Thus, the CNOT is not a simple gate and can produce a complicated result that depends upon the qubit on which it operates. (One might be tempted to put this down to the use of θ rather than $\theta/2$ in the arguments of Sect. 2.2, but these results would remain peculiar.)

Quantum circuits are obtained by utilizing a series of gates arranged in a particular order which depends upon the algorithm being implemented. While gates need to have universal properties (see the next sub-section), it is not clear that circuits need to have these properties, since the actual circuit implementation may well depend upon the technology at hand. Nevertheless, it is easy to set out a circuit in terms of the gates available, and at this point the circuit is likely to be independent of the technology. Some typical higher-order gate circuits, including the quantum equivalent of a classical full adder circuit have been discussed in the useful treatise [28].

In recent years, some approaches to quantum information have moved beyond simple descriptions in terms of qubits. Here, the concept of a *qudit* has appeared, and this quantity is described as a quantum version of d binary digits with $d > 2$ [29]. It is generally felt that using this higher-dimensional state space provides greater opportunity to describe algorithmic operations more complex than that afforded by the simple qubits above. Whether or not this is true is probably only a point of view, but programmable quantum processors have been developed with qudit approaches [30]. (Perhaps, a comparison can be made with the classical case in which designers talk about registers and full adders being the elements of interest, when in fact each is made up of a set of simple gates.) In general, though, the translation from a desired algorithm or program to acquiring the computational output is not conceptually different for a quantum computer than for a classical computer, as depicted in Fig. 2.5. First, the user finds an algorithm, or a program, to compute the quantity of interest. This may be done in a variety of programming languages. This program is then compiled onto the processor (using a compiler language that is usually unique to the programming language used). This step classically generates machine language suitable to the hardware implemented in the machine. In the quantum world, this is compiled into sets of quantum gates, either qubit or qudit, that work with the technology in hand. These instructions are then implemented on the hardware; classically, the microprocessor has a program control unit that translates machine language into the proper sequence of gate operations. The computation is then done and the output prepared. In the quantum world, many companies have generated programming software that aids in translating the program into operations on the quantum computer. One such set is the open source Qiskit programming environment [31], that is adaptable to many of the available quantum computers.

The 7 Requirements for Qubits. At the end of the twentieth century, David DiVincenzo, who was working at IBM Research at the time, published what he considered to be seven requirements for qubits in quantum information [32]. Since then, these conditions have become embedded within the community as critical criteria that must be met. At the same time, they have become guides to the development of quantum information. These seven requirements can be stated as:

Fig. 2.5 The general flow from program to data output in a computational scheme, whether classical or quantum

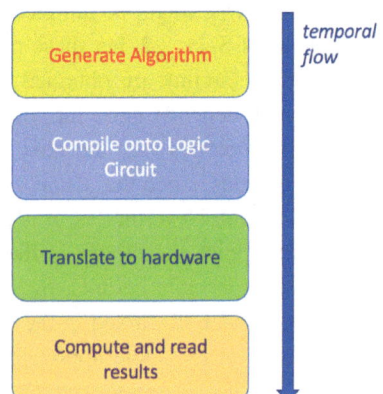

1. A scalable physical system with a well-characterized qubit,
2. The ability to initialize the qubit,
3. Long relevant coherence times,
4. A "universal" set of quantum gates,
5. An ability to measure the state of a qubit.

These first five apply to all realms of quantum information, whereas the next two are more specific to quantum communications involving:

6. The ability to convert a flying qubit to a stationary qubit, and vice versa,
7. The ability to faithfully transmit qubits from one point to another.

In general, these requirements might seem to be obvious today, but it was clear in the early days that they needed to be pointed out, if for no other reason than to clear away some of the not so obvious approaches. Scalability is one of the more obvious requirements, if only in comparison to today's nanoelectronics. The latter has progressed through the scalability of the microchip, and if quantum computing is to proceed to millions of qubits, the technology must be scalable. Were this not the case, the needed space and long communication time would eventually doom the project. Similarly, the need for a well characterized qubit is relatively obvious, as the lack of such a qubit would imply a failure to predict the performance of the system.

We have already pointed out the need for long coherence times. If these were not obtainable, then it would be exceedingly difficult to perform qubit operations and complete a calculation before all information would be lost. This need has little to do with any requirements on dissipation or entropy; it is simply the requirement that the data be reliable and not be lost during the computation. The lack of sufficiently long coherence times, and fully reliable gates brings the need for error correction techniques.

In the above, we have pointed out a variety of single-qubit and two-qubit gates. Whether or not this is a sufficiently robust set of qubit operations is open to discussion. However, these operations do not depend upon the actual physical implementation of the qubit, so they may be considered at least to be universal at some level. Classically, nearly all complicated logical operations can be reduced to sets of simple operations, so it may be assumed that the same is true with qubit operations. So far, this seems to be a reliable assumption.

The ability to measure the state of a qubit lies at the heart of quantum sensing, whether it is in a quantum computer or a quantum metrology system. This goes to the heart of the Turing principles discussed in Chap. 1, where the answer to the computing algorithm must finally be readable. If the processor, and its program, do not yield a stop state with the result in a readable condition, it is hard to consider this as a real computer.

Quantum communications actually takes qubits that are prepared in a desired state, encrypted, and then converted to flying qubits needed to carry the information from one site to the other. At the receiving end they have to be "caught" and then read into another set of stationary qubits which can be measured in order to decrypt them and find the signal. This lies at the heart of points 6 and 7 above.

2.5   Qubits in Communications

Quantum information networks are currently one of the biggest topics in quantum communications, and at the core lies the teleportation problem [33]. But, there is a problem and that is in quantum measurements. With a discussion of communications, our attention is once more turned to the process of measurement. To begin this, the polarization of a photon is considered. Photons themselves have an integer spin, usually denoted by ± 1, where the upper sign denotes a right-circularly polarized wave, and the lower sign denotes a left-circularly polarized wave. That is, the polarization which usually denotes the electric field direction of the electromagnetic wave for the right-circular polarized wave rotates in a right-hand sense as it propagates along a direction denoted by the extended thumb, and vice versa. Normally, optical scientists tend to denote the polarization as a combination of these two results in either vertical or horizontal polarization. In this case, an optical qubit can be described by one of the following states for the qubit:

$$
\begin{aligned}
|\uparrow\rangle &= |0\rangle \\
|\leftarrow\rangle &= |1\rangle \\
|\nearrow\rangle &= \tfrac{1}{2}(|0\rangle + |1\rangle) \\
|\searrow\rangle &= \tfrac{1}{2}(|0\rangle - |1\rangle)
\end{aligned}
\qquad (2.69)
$$

Obviously, the last two are mixtures of the vertical and horizontal polarizations. The reader should be reminded that we can tell the difference between any two states in which a qubit may exist only if they are described in the same coordinate system with the same basis set. Generally, qubits propagate via unitary transformations, such as the matrix operations described above for qubit gates. The question of interest here is just how to measure the gates.

As discussed in Chap. 1, the measurement problem is as old as quantum mechanics itself. And the crux of the problem lies immediately within the interpretation of quantum mechanics that one chooses to believe [34]. The trouble with measurements is described further elsewhere [35]. Here, we do not delve into some interpretations that are confusing and do not square with current measurements on quantum systems. Nevertheless, several things may be said about the measurement on a quantum system:

1. The measurement must yield a result that exists in a classical state. Contrary to some interpretations, here it is assumed that the quantum state is not destroyed, but may be altered in this process [36].

2. The apparatus of a measurement introduces its own set of basis functions that describe the possible outcomes of the measurement, and this outcome projects the quantum state onto the classical basis states.
3. Crucially, the measurement requires that \ the quantum system be opened to the environment, and this may well modify the quantum system and allow dissipative processes to affect the system [36].

In spite of these rather stringent requirements, there are two systems in which it has always been claimed that measurements can be made. These are spin systems in which measurements such as the Stern-Gerlach method are useful [37], and photons in which only the phase must be determined. On the other hand, the plethora of experiments being made today which leave the quantum structure intact, suggests that these two experiments are not alone in their usefulness.

Teleportation. To proceed to multiple qubits, it is going to be convenient to use the second form of the notation in (2.59) in which the control signal is included in the state matrix, so that this last equation becomes

$$\begin{bmatrix} 0a \\ 0b \\ 1a \\ 1b \end{bmatrix}, \tag{2.70}$$

where the first number obviously is the state of the control bit (or a second qubit). Thus, the operation of the CNOT gate would now be written as the change

$$\begin{aligned} |00\rangle &\to |00\rangle \\ |01\rangle &\to |01\rangle \\ |10\rangle &\to |11\rangle \\ |11\rangle &\to |10\rangle \end{aligned} \tag{2.71}$$

Another important consideration is that one cannot copy a qubit, as the qubit is a quantum state, and quantum states cannot be copied. This goes along with what is known as a no-cloning theorem–one cannot clone a copy of a state. This arises from the fact that all states in a quantum system are unique, so there can be no copy of a given state (the act of quantization of a system leads to a set of parameters, known as quantum numbers, and each available quantum state in the system has a unique set of these numbers–no two states can have the same set of numbers).

The general problem of communication can be expressed by the common argument that appears in many textbooks. Consider two people, Alice and Bob (these are the usual suspects in quantum communications, sometimes supplemented with Carol), who are presented upon their union with a pair of entangled photons, which are the entanglement of two particles, cleverly denoted 1 and 2. At a subsequent time, Bob travels to a great distance leaving Alice at home. Bob takes particle 2 with him, while particle 1 stays at home with Alice. Now, at this time, Carol appears on the scene, and asks Alice to send a secret message, encoded on particle 3, to Bob. (One could imagine the various reasons for this and the underlying sociology, but that is not the point here.) Alice cannot just directly send particle 3 to Bob, because of the various risks of sending information over the usual open air or fiber communication systems. With quantum communications, the approach taken is for Alice is to perform a joint measurement between her particle 1 and Carol's particle 3. Then, she needs only send a classical communication to Bob describing the results of this measurement. Upon receiving Alices' measurement result, he now makes a set of four measurements upon his own particle 2. These four measurements use the I, X, Y, and Z gates described above. He actually has to do only one of these, corresponding to the result sent to him by Alice. This measurement produces the equivalent particle 3 (which was destroyed in Alice's measurement, so there is no cloning here).

Suppose that the particles here are photon states. The entangled state is that considered already in (2.1), but in terms of the photon states for vertical and horizontal polarization. Thus, the original entangled state between particles 1 and 2 may be written as

$$|\psi\rangle_{12} = \frac{1}{\sqrt{2}}\big[|\rightarrow\rangle_1|\uparrow\rangle_2 - |\uparrow\rangle_1|\rightarrow\rangle_2\big]. \tag{2.72}$$

When Alice makes the joint measurement of particles 1 and 3, she entangles them into the state

$$|\psi\rangle_{13} = \frac{1}{\sqrt{2}}\big[|\rightarrow\rangle_1|\uparrow\rangle_3 - |\uparrow\rangle_1|\rightarrow\rangle_3\big]. \tag{2.73}$$

Then, when her measurement is made by measuring the polarization of particle 3, this fixes the polarization of particle 1. She then sends to Bob the result of her measurement and the manner in which particles 1 and 3 were entangled. Bob can now use this information, including on whether to use I, X, Y, or Z as a transform, on his particle 2, and the result correlates to the polarization of particle 3 [28].

Encryption and Communications. The second classical example is that Alice needs to send to Bob, for whatever reason, the local weather at her home. Before leaving, Bob and Alice decided to use two qubits to encode the weather in the following form:

$$
\begin{aligned}
|00\rangle &\rightarrow freezing, clear \\
|01\rangle &\rightarrow freezing, cloudy \\
|10\rangle &\rightarrow non-freezing, clear \\
|11\rangle &\rightarrow non-freezing, cloudy
\end{aligned}
\tag{2.74}
$$

Using entangled qubits as before will allow Alice to send this information using a single qubit; e.g., sending two qubits of information via a single qubit. This is known as dense encoding and arises from entanglement. While dense coding is known in classical data, the use of quantum approaches allows a denser encoding. To begin, Alice and Bob create a set of entangled particles as before, but now use a positive sign (relating to a higher energy level of the atom, when atoms are used, but this allows two identical particles to be created in what is called the triplet state, as this level has 3 available distinct states) so that

$$|\psi_{00}\rangle = \frac{1}{\sqrt{2}}(|00\rangle + |11\rangle) = \frac{1}{\sqrt{2}}\begin{bmatrix} 1 \\ 0 \\ 0 \\ 1 \end{bmatrix}, \tag{2.75}$$

where the earlier notation is used. Bob takes one of the entangled particles with him on his journey, while Alice retains her particle. To encode the weather state, Alice prepares a qubit to be sent, using a different transform for each state of the weather.

What Alice is going to do is to encode (or encrypt) the above qubit using the state of the weather. For each weather state, she will use a different gate transformation on the qubit (2.74). She will then send this encoded qubit to Bob, where he will use his entangled particle to decode the signal. Alice choses to use the simplest gates, which will be the I, X, Y, Z gates, and the choice of gate depends upon the weather. For the weather state $|00\rangle$, she uses the identity matrix, which leaves the qubit (2.73) unchanged, as

$$|\psi\rangle_{00} = \frac{1}{\sqrt{2}}(|00\rangle + |11\rangle) = \frac{1}{\sqrt{2}}\begin{bmatrix} 1 \\ 0 \\ 0 \\ 1 \end{bmatrix}. \tag{2.76}$$

For the second weather state $|01\rangle$, she decides to use the Z transform which is applied only to the first value in each of the pairs in (2.73). This multiples the first zero by 1, and the second (1 in this case) is multiplied by -1. This converts (2.73) to

$$|\psi\rangle_{01} = \frac{1}{\sqrt{2}}(|00\rangle - |11\rangle) = \frac{1}{\sqrt{2}}\begin{bmatrix} 1 \\ 0 \\ 0 \\ -1 \end{bmatrix}. \tag{2.77}$$

So far, this is easy enough. For the third weather state $|10\rangle$, she decides to use the X transform, again to the first term. This transform takes the form

$$X\psi = \begin{bmatrix} 0 & 1 \\ 1 & 0 \end{bmatrix}\begin{bmatrix} 0 \\ 1 \end{bmatrix} = \begin{bmatrix} 1 \\ 0 \end{bmatrix}, \tag{2.78}$$

where (2.52) has been used so that the encoded qubit (2.73) now becomes

$$|\psi\rangle_{10} = \frac{1}{\sqrt{2}}(|10\rangle + |01\rangle) = \frac{1}{\sqrt{2}}\begin{bmatrix} 0 \\ 1 \\ 1 \\ 0 \end{bmatrix}. \tag{2.79}$$

Finally, for the fourth weather state $|11\rangle$, she uses the $-iY$ transform, that takes the form of

$$-iY\psi = -i\begin{bmatrix} 0 & -i \\ i & 0 \end{bmatrix}\begin{bmatrix} 0 \\ 1 \end{bmatrix} = \begin{bmatrix} -1 \\ 0 \end{bmatrix}, \tag{2.80}$$

where (2.53) has been used. Now, the encoded qubit is given as

$$|\psi\rangle_{11} = \frac{1}{\sqrt{2}}(-|10\rangle + |01\rangle) = \frac{1}{\sqrt{2}}\begin{bmatrix} 0 \\ 1 \\ -1 \\ 0 \end{bmatrix}. \tag{2.81}$$

Depending upon the weather, Alices sends one of these four qubits to Bob.

Upon receiving the qubit from Alice, Bob now has to decode (or decrypt) the information in Alice's signal. Bob's first step is to apply the CNOT operation (2.62) to the received qubit, which gives the results

$$T_{CNOT}|\psi\rangle_{00} = \frac{1}{\sqrt{2}}\begin{bmatrix} 1 \\ 0 \\ 1 \\ 0 \end{bmatrix} \quad T_{CNOT}|\psi\rangle_{01} = \frac{1}{\sqrt{2}}\begin{bmatrix} 1 \\ 0 \\ -1 \\ 0 \end{bmatrix}$$

$$T_{CNOT}|\psi\rangle_{10} = \frac{1}{\sqrt{2}}\begin{bmatrix} 0 \\ 1 \\ 0 \\ 1 \end{bmatrix} \quad T_{CNOT}|\psi\rangle_{11} = \frac{1}{\sqrt{2}}\begin{bmatrix} 0 \\ 1 \\ 0 \\ -1 \end{bmatrix} \tag{2.82}$$

which merely interchanges the last two values in each qubit representation. In the representation of the entangled pair (the first formulation in, for example, (2.73)), these new states (distinguished by the primes) can be written as

$$|\psi'\rangle_{00} = \tfrac{1}{\sqrt{2}}(|00\rangle + |10\rangle) \quad |\psi'\rangle_{01} = \tfrac{1}{\sqrt{2}}(|00\rangle - |10\rangle)$$
$$|\psi'\rangle_{10} = \tfrac{1}{\sqrt{2}}(|10\rangle + |11\rangle) \quad |\psi'\rangle_{11} = \tfrac{1}{\sqrt{2}}(|10\rangle - |11\rangle) \tag{2.83}$$

Notice that the last digit, $|0\rangle$ in the case of the first row and $|1\rangle$ in the case of the second row, can be factored out. A measurement of this second bit separates the possible results into two pairs of bits, so that one can proceed with just one of the two pairs, depending upon the results of the measurements. This measurement does not affect the first bit in any way. Bob now applies the Hadamard transform to the four qubits, operating only upon the first state bit. This gives the results (the double prime represents the states after this last transform)

$$
|\psi''\rangle_{00} = \frac{1}{2}\begin{bmatrix} 1 & 1 \\ 1 & -1 \end{bmatrix}\begin{bmatrix} 0 \\ 1 \end{bmatrix} = \begin{bmatrix} 1 \\ 0 \end{bmatrix} \quad |\psi''\rangle_{01} = \frac{1}{2}\begin{bmatrix} 1 & 1 \\ 1 & -1 \end{bmatrix}\begin{bmatrix} 0 \\ -1 \end{bmatrix} = \begin{bmatrix} 0 \\ 1 \end{bmatrix},
$$
$$
|\psi''\rangle_{10} = \frac{1}{2}\begin{bmatrix} 1 & 1 \\ 1 & -1 \end{bmatrix}\begin{bmatrix} 1 \\ 1 \end{bmatrix} = \begin{bmatrix} 1 \\ 0 \end{bmatrix} \quad |\psi''\rangle_{11} = \frac{1}{2}\begin{bmatrix} 1 & 1 \\ 1 & -1 \end{bmatrix}\begin{bmatrix} 1 \\ -1 \end{bmatrix} = \begin{bmatrix} 0 \\ 1 \end{bmatrix},
$$
$$(2.84)$$

where we have used (2.58). Now, the previous results were separated as either row 1 or row 2 by the measurement of the second bit. Now a measurement of the results of this Hadamard transform will separate the results as column 1 or column 2, and the result of the two measurements determines entirely which of the four states was sent.

Bob was able to decode Alice's signal only because he had the "key", the set of transforms that he needed to apply to the qubit sent to him by Alice. In this case, we may assume that Bob possessed the key because he knew the set of operations that would be applied by Alice. In more complicated cases, the recipient of the message needs to know the key, and this may need to be distributed from the source of the message. Key distribution will be discussed in the next section. Problems that arise in quantum communications are many. While much is known about this area, whether or not quantum communications is yet in the main stream remains to be determined. Adoption is still a point of discussion as to when and how [38]. Then, how the quantum data is to be transported (or teleported) can be done in many ways, and there is no common standard as yet, although some simple approaches exist [39–41]. Two points are critical here: the use of quantum resources must be optimized [42], and there must be a form of time synchronization in the system [43], as discussed earlier.

Before proceeding further, it is probably worthwhile to mention a little bit of classical encryption that is used today. The Advanced Encryption Standard that has been adopted by the National Institute of Standards and Technology (NIST) depends upon a substitution-permutation network, and when the key is 256 bits, the result is almost unbreakable [44, 45]. Other approaches have been developed using noise to encode the data, with results that are said to be as capable as quantum encryption [46, 47]. However, much is known, but not spoken about in the open literature, in regard to the state of cryptography.

Quantum Key Distribution. Today, there are many different methods of quantum key distribution and many methods of encrypting the information onto the set of qubits that are to be communicated. One can think of many necessary steps that have to be achieved toward achieving useful quantum communications. For example, it is relatively obvious that the channel, and the key distribution, should be independent of the qubit technology that is used [48]. In addition, the distribution of keys should be relatively rapid, in order not to hinder the communication channel [49, 50]. In many cases, with optical communications, this will require single-photon detectors [51]. Secondly, the key distribution should be safe and secure in its process, and this requires some system optimization and likely error correction [52]. In many cases, a continuous distribution of (changing) keys is required for more security, and this also requires high-speed processing [53].

Having said this, what constitutes a good code is open to interpretation. And, it is quite likely, that much of the discussion is not conducted in the open community. Nevertheless, there are some protocols that have made it into the open literature. One such is the BB84 protocol, developed by Bennett and Brassard [54]. As in most key distribution approaches, a signal is sent from Alice to Bob that is encrypted much like the weather example above, and Bob obtains the key for future signals by decrypting this signal from Alice. The BB84 approach has been shown to be relatively robust against imperfections [55]. Another more modern approach is based upon classical encryption methods based upon random noise, perhaps generated from chaotic systems, and this approach has moved to the quantum sector [56].

Another common approach is the use of polar codes [56]. These codes are said to achieve the maximum channel capacity possible by the concept of channel polarization [57]. In this approach a multiple recursive operation concatenates information onto a short kernel code, which transforms the single channel into a set of virtual "outer" channels. When the number of recursions is large, the channel can become nearly noiseless, and it satisfies the Shannon limit (roughly an energy of $k_B T ln(2)$ per bit) on information transmission [58]. Such codes are adaptable to continuous variable quantum key distribution as well [59]. As mentioned in the BB84 above, in this approach, the initial process is the generation of an encrypted signal that contains an initial key and this is sent to Alice and Bob. They post process this information, and use a classical channel that contains some properties they need. Yet, their results are likely not the same due to many factors, so they have to also perform a reconciliation process with error correction to obtain consistent key information. These keys are then used to decode subsequent information on new keys, if necessary.

Networks and Elements. Creating quantum networks is a problem at a variety of levels. First, there is the overall strategy for developing a network [33]. This latter is a global view. At the microscale view, there is the problem of converting the local qubit to a flying qubit, whether this flying qubit is at the high microwave frequency or optical. And, there is the reverse problem of converting the flying qubit to a local qubit for processing at the receiving end. In between, there is the problem of routing the flying qubits from the source to the reception end. In classical communication networks, the routing problem is contained within processors that are known as routers, and there are usually a multitude of these between the observer's computer and the source of the web page that is being observed. Even at a simple home, the cable modem is also a router that sends wifi signals to all the computers, iPads, and phones that are seeking data, and keeps track of what is being sent where–e.g., of making the proper connection to the phone, iPad, or computer asking for the data.

Needless to say, there is a large community that is addressing this multitude of problems (only a small sample of the work is mentioned here). In the central part of the communications channel, work is addressing routers [60] and switches, that can rearrange parts of the channel [61]. The conversion from a microwave local qubit to an optical flying qubit has been addressed, for example with an Er ion intermediary [62]. However, it is also possible to take the qubit state and use it to drive solid-state single-photon sources to create directly the optical signal [63]. It has also been shown that it is easily possible to change from a picosecond photon to a nanosecond photon, which would be for using optical processing at the receiving end of the channel [64]. And, the conversion from a telecom qubit to a solid-state qubit has been demonstrated [65]. While these are some recent examples for the state-of-the-art, the field is much richer, and can be traced backwards from these examples.

2.6   Quantum Sensing

There is a distinct view in the community that quantum sensing can improve the ability of measuring systems by employing quantum elements [66]. For this, a variety of quantum processes have been explored, both at the qubit level and extending to the use of critical phenomena such as phase transitions [67]. While quantum sensing is thought to be a relatively recent field of study, quantum systems have been used for metrology for quite some time. Examples are compact atomic clocks for time measurements and superconducting quantum interference devices, utilizing Josephson junctions, the latter of which also are used for qubits (to be discussed in Chap. 3). In sensing, one seeks to do one of three things [68]: (1) measure a physical quantity with a quantum object, (2) use quantum coherence to enhance the measurement of a physical quantity, or (3) use entanglement to improve the sensitivity of a measurement. In each of these cases, the object is to overcome classical limits in the measurement of the physical quantity of interest, whether it is time or the result of a measuring apparatus. In addition, quantum sensing is subject to many of the DiVincenzo criteria listed above [32], but also must satisfy the requirement that the system to be measured must interact with the measurement tool in a manner that involves a relevant physical quantity such as charge, flux, voltage, current, temperature, etc. [69].

In the classical world, measurements are always limited in some way, and this way is usually the noise that is always present. The measurement problem is to sense the desired signal in a background of noise, and then to make a measurement of this signal. There are normally considered to be two major sources of noise in electronic systems. The first, and most common, is thermal noise, often called Johnson noise or Johnson-Nyquist noise, after the discoverer [70] and his colleague who explained it [71]. Generally, this noise is relatively broad-band and has a noise power given by

$$k_B T \, \Delta f, \tag{2.85}$$

where Δf is the band-width of the detector/receiver, T is the temperature and k_B is Boltzmann's constant. Often, the noise is expressed as a root-mean-square value of the noise voltage or noise current. Perhaps the most common experimental method of detecting weak signals is the use of synchronous detection, which usually involves a so-called lock-in amplifier, and is a method to reduce Δf by a significant amount. This approach typically modulates the incoming signal by a low frequency, and this searches for this modulation signal in the output of the detection system (this basically assumes the main source of noise is the detection system itself) [72]. While it might be thought optical signals can avoid this, one only has to recall that the detectors will have this noise.

The second type of noise is known as $1/f$ noise and generally has a spectrum that varies as $1/f^\alpha$, with α varying $0 < \alpha < 2$. This noise is often the dominant noise in biological systems, but it is present in many devices in the microwave and optical regions of the spectrum as well [73]. In condensed matter, this noise is thought to arise from defects and traps that have a wide range of excitation energies. Sometimes it is referred to as "pink" noise, since the spectrum is has more energy at the lower frequencies (toward the red, or pink, part of the spectrum for optical signals).

When one seeks to find detectors/receivers with less noise, it is believed that quantum sensors can be used to achieve lower noise results. This means that the fixed energies of atomic energy levels, or the use of photons of fixed energy will result in the development of improved low-noise detectors. This is the world of quantum sensing [74]. But, as with the use of the lock-in amplifier, lower noise performance can be achieved only with the use of additional resources. One argument is that enhanced performance proportional to $1/\sqrt{N}$, where N is the sum of the resources utilized, can be achieved classically. This can be the number of different simultaneous (or sequential) measurements or the number of parallel channels studying the quantity of interest, or even the number of people viewing the apparatus. The view is that the use of quantum resources can improve this to π/N, which is known as a form of the Heisenberg limit [75]. But, to achieve this limit requires an efficient utilization of all the available resources [76]. While resources are important, there are actually estimates of the expected accuracy of a measurement in the presence of noise even in the quantum regime. It must be recalled that the measurement projects the quantum system onto a set of classical results determined by the measurement system, so that, at the end, a classical result is obtained.

A classical limit on communication channels addressed the problem of measuring/sensing the signal. Shannon postulated that if the power in the signal was P, and the noise power was N, then the maximum bit rate that could be used is [77]

$$C = \Delta f \log_2 \left(\frac{P+N}{N} \right). \tag{2.86}$$

Several people have furthered this to say that this requires an energy per measurement is the limit provided by Landauer in erasing a bit [78]:

$$E = k_B T ln(2). \tag{2.87}$$

This approach has led to setting limits for quantum sensing, approaching this from the same estimation theory. One approach is based upon the Fisher information F_C, which may be determined for a fixed quantum state, and a measurement M of the parameter ϑ, as [79]

$$F_C[M] = \sum_i Tr\{\rho_\vartheta M\} l_i^2, \tag{2.88}$$

where ρ_ϑ is the quantum state corresponding to the observable ρ_ϑ and

$$l_i = \frac{1}{Tr\{\rho_\vartheta M\}} Tr\left\{ \frac{\partial \rho_\vartheta}{\partial \vartheta} M \right\}. \tag{2.89}$$

This means that the expected figure-of-merit depends not only on the quantum state (and the quantum measurement system), its variation with the parameter, but also upon the details of the measurement itself. In estimation, ϑ is considered to be a continuous variable; when it consists of only a discrete number of finite values (such as in the case of digital information), it is said to be a discrimination problem [80]. Equation (2.89) discusses the former case of a continuous valued parameter. The form of this equation is one of several that address Fisher information. In the case of discrimination, one has to synchronize the measurements with the data flow, and this typically uses a storage register for a set of the data. Operations can be made on this data, but evaluation of the limits still is concerned with the Fisher information [80]. Nevertheless, there has been progress in experimental measurements in which quantum-enhanced receivers have provided better performance for classical communications [81], and it has been shown that enhanced quantum control can affect the noise performance as well [82]. This couples with other studies that tend to show that quantum control provides an improvement in the ability to go beyond classical limits [83].

Limits on Energy. In the discussion about Eqs. (2.86) and (2.87), it was remarked that there is likely a minimum energy required to read a bit of information, whether this is a classical bit or a qubit. While the Landuaer limit (2.87) is for the erasure of a bit, it has long been taken to be the limit on reading a bit. He argued that the degrees of freedom associated with the information contained in 2^N states associated with the N bits in the machine would be coupled to an entropy increase of $k_B N ln(2)$. When this information was erased, or lost, the result would be the energy dissipation [84]. It was further argued that this dissipation would result in additional noise in the system, but additional accuracy could be obtained by computing slowly. Actually, (2.86) is not unique to Landauer, as it is often attributed to Shannon's work on communication theory as the necessary energy to read a transmitted bit. While (2.86) does not appear explicitly in [77], it is certainly obtainable for a binary stream.

How (2.87) would lead to (2.86) when the signal power was less than the noise power was shown by Keyes [85]. Originally, Shannon was concerned with information in communication systems, and referred to any missing information as the signal entropy [86]. This connection of information to negative (physical) entropy has continued up to the present, although they are not equal in most cases [87, 88]. Once Shannon had coined his information entropy, Brillouin worked out in some detail the connection of this to physical entropy [89]. But, these quantities are not the same thing, and do not yield the same numerical values in most cases (entropy will be discussed further below).

The question of the role played by degeneracy in condensed matter systems was examined by Bate [90], who examined the energy required to switch a bit in the presence of the quantum statistics. In spite of the differences in the statistics of degenerate systems, he again obtained (2.86). In doing so, he strengthened the argument that (2.87) also is quite likely to apply to quantum systems. Others have examined more closely classical computing systems and set a variety of other limiting factors [85, 91, 92]. While interesting, it is not yet clear whether any, or all, of these actually applies to quantum information processing, although some authors have undertaken this task [92, 93].

Entropy. The concept of "entropy" is thrown around extensively in the computer literature, but it does not seem to be understood well. Physical entropy and information entropy mean totally different things [87, 88]. It is important to clearly distinguish between these two quantities that have two distinct meanings and that cannot be used interchangeably. In Chap. 1, the idea of information-bearing states and non-information-bearing states was introduced. The differences in the two versions of entropy have a close connection to these two concepts. As an example, in terms of recent arguments above, the information-bearing states can be the 2^N possible configuration of the N bits or qubits in the system. But, these are macro-states! That is, they are macroscopic states that are created by the technology of the bits or qubits. In the condensed matter system, in which the bit or qubit is created, there are a great many more micro-states consisting, for example, of all the various electrons that exist in the microchip in which the bit or qubit is embedded. All of the micro-states are included in the non-information bearing states of the system. Historically, Boltzmann defined entropy as being a property of a macro-state, for which the entropy S is related to the number of microscopic configurations, W, that are compatible with that macro-state. This is expressed as [94, 95]

$$S = k_B ln(W), \qquad (2.90)$$

where k_B is the Boltzmann constant. Now, consider an example in which N particles can exist in a number of micro-states and each of these micro-states has a probability p_i of actually occurring. It can be shown that the entropy of these N particles is then given by [96]

$$S = -k_B \sum_i p_i ln(p_i). \qquad (2.91)$$

This last equation has great similarity to one used to express information entropy in computing or other applications, which is why it is given here. The problem is exactly that discussed above. Physical entropy is a result of studies of the micro-states that make up any given macro-state. Information entropy studies the macro-states that represent bits or qubits, and this involves only the information-bearing states of the system, not the much larger set of non-information-bearing states. While the physical entropy, or equivalently the thermodynamic entropy is linked to the energy in the system, there is no connection between information entropy and energy. These two entropies are like apples and oranges [88].

Semiconductor Sensing. Perhaps the most common sensor arises from the use of semiconductors and semiconductor devices. These are common particularly in the optical and microwave regimes. But, for quantum sensing, most effort has focused on the single-electron transistor. The basics of these devices will be discussed in Chaps. 3 and 5, and the semiconductor forms will appear in various qubits and sensing circuits to be described in the latter chapter.

Optical Sensing. Nearly all cable communications, and a great many over-the-air communication technologies are based on optical signals. In the world of quantum optical signals, and quantum sensing techniques, two important parts are single photon sources and single photon detectors [97]. Single photon emitters have come of age and there are several scalable approaches available for these [98]. The conversion of microwave signals to optical signals can be done in many ways, for example using the unique properties of graphene [99], but this does not produce single photons. The use of single-photon sources is an enabler for the generation of non-classical light which then becomes useful in quantum metrology.

In metrology, it is well-known that classical light can be used with interferometers for phase sensitive measurements. The use of weak non-classical light can greatly improve the sensitivity of the measurement, often by more than an order of magnitude [100]. Here, the nonclassical light can be a squeezed state, which can involve therefore single photons. In most instances, the generation of the nonclassical light involves processing in integrated photonic systems [97]. Consider the problem described in Fig. 2.6, in which a number of phase shifts are to be measured. The phase shifts are the parameters ϑ_i, which are distributed over the m input modes to the measuring system. The operators $\hat{a}_i$ are the mode inputs, and $\hat{b}_i$ are the outputs for which the phase is to be measured. Here, only a single mode is to be replaced by the nonclassical photons represented by the quantum state $\hat{\rho}$, in keeping with the ideas of (2.89). In the figure, it is suggested that the nonclassical light may actually be amplified. This can occur even in a unidirectional fiber through which the photons are transmitted [101, 102]. By introducing the non-classical light, the precision of the measurement (inverse of the minimum error in the measurement) can be

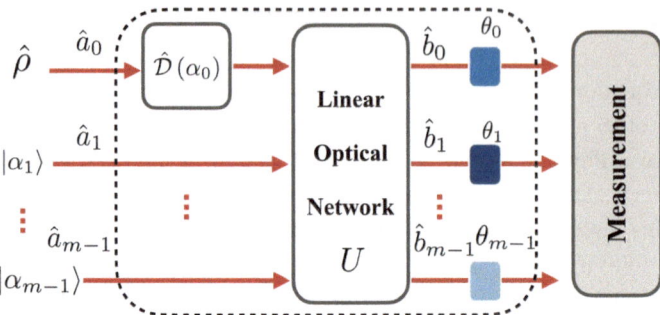

Fig. 2.6 A distributed quantum sensing scheme consists of a passive linear optical network, in which a single nonclassical state $\hat{\rho}$ is introduced. It may be displaced in amplitude as shown in the upper left block for the quantum channel. The different channels may be sent to different locations where the phase shift measurements are made. Reprinted from Ge et al. [100], under the creative commons CC By license

increased from $\sqrt{N_C}$ to $\sqrt{4N_QN_c}$, where N_C is the number of classical photons and N_Q is the number of non-classical photons.

One take on the multimode system of Fig. 2.6 lies in the world of pattern recognition, where the multiple modes are actually the pixel information captured in image processing [103]. In this work, a grey scale image of $P \times Q$ pixels, which is recorded in a matrix of the same dimension, is discussed. This matrix data is then fed to an m qubit system (with $2^m > N = P \times Q$). The information from the image is fed in column order into a register with the m qubits. Then, n of these qubits ($2^n = N$) are used to form a quantum state, on an arbitrary basis set $\{|i\rangle, i = 1, 2, ..., n\}$ as

$$|\psi\rangle = \sum_i c_i |i\rangle, \tag{2.92}$$

where the c_i are the pixel values. This information is fed to the quantum processor along with the equivalent data from a reference image; the purpose in this example was recognition of an image of a digital number, so the reference was another number. The processing can refer to a number of tasks: one being the so-called Hamming distance between the images (in two sequences of bits, this is the number of bits that differ between the two strings), the edges of the image, and so on.

At the other end from the quantum source lies the detectors, such as the image processing system of the previous paragraph, where the actual detector(s) read the image information into the system as the values c_i used in (2.92), whether this was a grey scale or a color image. It has already been pointed out that quantum effects can enhance classical receivers of almost any type [81]. The fibers discussed above can even be used as the detector, especially for some physical quantities as temperature [104].

In Sect. 2.3, the parity operator (2.45) was introduced. This parity operator is connected to the Wigner function which is often used in quantum optics to display the nonclassical nature of the states [105, 106]. The Wigner function uses the density matrix (2.11) of two spatially varying wave functions, each displaced from their common position centroid, and then taking the Fourier transform of the separation as the momentum [107]. This produces a quasi-probability function in position and momentum space, which can then be compared to a classical form. The non-classical Wigner function arises when it has negative excursions (which do not occur in classical statistics). The Wigner form of the parity operator uses two displacement operators $\widehat{D}(\alpha)$, where α is the displacement [17], and has the form [108]

$$W(\alpha) = \frac{2}{\pi} \langle \widehat{D}(\alpha) \widehat{\Pi} \widehat{D}^{\dagger}(\alpha) \rangle. \tag{2.93}$$

It has been shown to be especially useful in optical sensing of atom spectroscopy. Continuous variable states of known parity can be generated from single mode squeezed photonic states, and these can be especially useful in metrology [109].

Beyond Optical. In the microwave range, quantum sensing can greatly improve the ability to measure signals. The implementation of a joint measurement using a superconducting resonator along with a microwave radar has demonstrated an improvement in target detection [110]. In this approach, the superconducting circuit provides an idler circuit to enhance the detection. Although, this was done within a dilution refrigerator, working at temperatures well below 1 K, it is hoped the results will be applicable to real radar systems. This result is consistent with the general concept of using cryogenic quantum sensors to enhance microwave power measurements [111].

The above use of superconducting circuits is a natural extension of the use of these items in general metrology. For example, the superconducting circuit can be a interferometer that generates entanglement [112]. When this circuit composes a qubit, it has been shown that enough sensitivity is obtained to measure neuronic signals [113]. These superconducting sensors will be discussed more fully in Chap. 3.

Another approach to quantum sensing is to use spin qubits that arise from the nitrogen vacancy in diamond (the so-called NV center). It has been shown that the equivalent of lock-in sensitivity can be obtained using such qubits and can be used with arbitrary frequency resolution [114]. These centers can be quite dense, which provides a strongly-interacting spin system [115]. In the case of pulsed systems, there can naturally be pulse width effects [116]. Nevertheless, the Quantum Fourier transform has been implemented with these qubits [117]. Finally, it should be remarked that there is a field of quantum sensing study that suggests building bottom-up qubits utilizing molecular systems [118]. The NV sensors will be discussed in Chap. 7.

References

1. Schrödinger, E.: Die gegenwartige Situation in der Quantenmechanik. Naturwiss. 23, 803, 827, 844 (1935); Tr. Trimmer, J. D.: The Present Situation in Quantum Mechanics: A Translation of Schrödinger's "Cat Paradox" Paper. Proc. Am. Phil. Soc. 124, 323 (1980).
2. Bohm, D., Aharonov, Y.: Discussion of Experimental Proof for the Paradox of Einstein, Rosen, Podolsky. Phys. Rev. 108, 1070 (1957).
3. Zeeman, P.: On the Influence of Magnetism on the Nature of Light Emitted by a Substance. Phil. Mag., Ser. 5, 43, 226 (1897).
4. Gribbon, J.: Erwin Schrödinger and the Quantum Revolution. (Wiley, New York, 2013).
5. Uhlenbeck, G. E., Goudsmit, S.: Ersetzung der Hypothese vom unmechanischen Zwang durch eine Forderung bezüglich des inneren Verhaltens jedes einzelnen Elektrons. Naturwiss. 13, 953 (1925).
6. Slater, J. C.: The Theory of Complex Spectra. Phys. Rev. 34, 1293 (1929).
7. Einstein, A.: Strahlungs-Emission und -Absorption nach der Quantentheorie. Verhand. Deutschen Phys. Gesell. 18, 318 (1916).
8. Ferry, D. K., Fannin, D. R.: Physical Electronics. (Addison-Wesley, Reading, 1971).
9. Rabi, I. I.: Space Quantization in a Gyrating Magnetic Field. Phys. Rev. 51, 652 (1937).
10. Morse, P. M., Feshbach, H.: Methods of Theoretical Physics. (McGraw-Hill, New York, 1953).
11. Jaynes, E. T., Cummings, F. W.: Comparison of Quantum and Semiclassical Radiation Theories with Applications to the Beam Maser. Proc. IEEE 51, 89 (1963).
12. Norozhny, N. B., Sanchez-Mondragon, I. I., Eberly, J. H.: Coherence versus Incoherence: Collapse and Revival in a Simple Quantum Model. Phys. Rev. A 23, 236 (1981).
13. Gea-Banacloche J.: Collapse and Revival of the State Vector in the Jaynes-Cummings Model: An Example of State Preparation by a Quantum Apparatus. Phys. Rev. Lett. 65, 3385 (1990).
14. Birrittella, R., Chang, K., Gerry, C. C.: Photon-Number Parity Oscillations in the Resonant Jaynes-Cummings Model. Optics Commun. 354, 286 (2015).
15. Hessian, H. A., Mohamed, A.-B. A.: Quasi-Probability Distribution Functions for a Single Trapped Ion Interacting with a Mixed Laser Field. Laser Phys. 18, 1217 (2008).
16. Shore, B. W., Knight, P. L.: The Jaynes-Cummings model. J. Mod. Optics 40, 1195 (1993).
17. Ferry, D. K.: Quantum Mechanics--An Introduction for Device Physicists and Electrical Engineers, 3rd Ed. (CRC Press, Boca Raton, 2021).
18. Merzbacher, E.: Quantum Mechanics, 2nd Ed. (Wiley, New York, 1970) Ch. 13; Feynman, R. P., Leighton, R. B., Sands, M.: The Feynman Lectures on Physics, Vol. III (Addison-Wesley, Reading, MA, 1965), Sec. 11.
19. Ashrafi S. M., Bazrafkan, M. R.: New Approach to Solving Master Equations of Density Operator for the Jaynes-Cummings Model with Cavity Damping. Chin. Phys. 23, 090303 (2014).
20. de los Santos-Sánchez, O., Récamier, J., Jáuregui, R.: Markovian Master Equation for Nonlinear Systems. Phys. Scrip. 90, 074018 (2015).
21. Abdalla, M. S., Khalil, E. M., Obada, A. S.-F., Perina, J., Krepelka, J.: Quantum statistical characteristics of the interaction between two two-level atoms and radiation field. Eur. Phys. J. Plus 130, 227 (2015).
22. Ghorbani, M., Faghihi, M. J., Safari, H.: Wigner Function and Entanglement Dynamics of a Two-Atom Two-Mode Nonlinear Jaynes-Cummings Model. J. Opt. Soc. Amer. B 34, 1884-1893 (2017).

23. Pandit, M., Das, S., Roy, S. S., Dhar, H. S., Sen, U.: Effects of Cavity-Cavity Interaction on the Entanglement Dynamics of a Generalized Double Jaynes-Cummings Model. J. Phys. B 51, 045501 (2018).

24. Faraji, E., Tavassoly, M. K.: Dynamics of Physical Properties of a Single-Mode Quantized Field Non-Linearly and Non-Resonantly Interacting with Two V-Type Three-Level Atoms Passing Consecutively Through a Cavity, Opt. Commun. 354, 333–343 (2015).

25. Lv, D., An, S., Um, M., Zhang, J., et al.: Reconstruction of the Jaynes-Cummings Field Sate of Ionic Motion in a Harmonic Trap. Phys. Rev. A 95, 043813 (2017).

26. Boole, G.: An Investigation of the Laws of Thought. (Walton and Maberly, London, 1854).

27. Goldstein, H.: Classical Mechanics. (Addison-Wesley, New York, 1951).

28. Marinescu D. C., Marinescu, G. M.: Lectures on Quantum Computing, un-published.

29. Wang, W., Hu, Z., Sanders, B. C., Kais, S.: Qudits and High Dimensional Quantum Computing. Fron. Phys. 8, 589504 (2022).

30. Chi, Y., Huang, J., Zhang, Z., Mao, J., et al.: A Programmable Qudit-Based Quantum Processor. Nature Commun. 13, 1166 (2022).

31. https://qiskit.org.

32. DiVincenzo, D. P.: The Physical Implementation of Quantum Computing. Fortschr. Phys. 48, 771 (2000).

33. de Forges de Parny, L., Alibart, O., Debaud, J., Gressani, S., et al.: Satellite Quantum Information Networks: Use Cases, Architecture, and Roadmap. Commun. Phys. 6, 12 (2023).

34. Oriols, X., Ferry, D. K.: Why Engineers are Right to Avoid the Quantum Re-ality Offered by the Orthodox Theory. Proc. IEEE 109, 955 (2021).

35. Ferry, D. K., Oriols, X., Weinbub, J.: Quantum Transport in Semiconductor Devices: Simulation with. Particles. (IOP Publishing, Bristol, in press).

36. Zurek, W. H.: Decoherence, Einselection, and the Quantum Origins of the Classical. Rev. Mod. Phys. 75, 715 (2003).

37. Gerlach, W., Stern, O.: Der Experimentalle Nachweis der Richtungsquantelung im Magnetfeld. Z. Phys. 9, 349 (1922).

38. Joseph, D., Misoczky, R., Manzano, M., Tricot, J., et al.: Transitioning Organizations to Post-Quantum Cryptography. Nature 605, 237 (2022).

39. Mastriani, M.: Non-Ambiguous and Simplified Quantum Teleportation Protocol. EPJ Quantum Technol. 10, 14 (2023).

40. Perepechaenko, M., Kuang, R.: Quantum Encryption of Superposition States with Quantum Permutation Pad in IBM Quantum Computers. EPJ Quantum Technol. 10, 7 (2023).

41. Liang, K., Cao, Z., Chen, X., Wang, L., et al.: A Quantum Secure Direct Communication Scheme Based on Intermediate Basis. Fron. Phys. 18, 51301 (2023).

42. Kumar, S.: Multi-Output Teleportation of Different Quantum Information with Optimal Quantum Resources. Opt. Quantum Electron. 55, 296 (2023).

43. Nande, S. S., Paul, M., Senk, S., Ulbricht, M., et al.: Quantum Enhanced Time Synchronization for Communication Network. Comp. Networks 229, 109772 (2023).

44. Announcing the Advanced Encryption Standard (AES). Fed. Inform. Proc. Stand. 197, NIST, November 26, 2001.

45. Biryukov, A., Khovratovich, D., Nikolic, I.: Distnguisher and Related-Key Attack on the Full AES-256. CRYPTO-2009. (Int. Assoc. Crypto. Res., 2009) pp. 231–49.

46. Kish, L. B., Granqvist, C. G.: On the Security of the Kirchoff-Law-Johnson-Noise (KLJN) Communicator. Quantum Info. Proc. 13, 2213 (2014).

47. Kish, L. B.: The Kish Cypher: The Story of KLJN for Unconditional Security. (World Scientific, Singapore, 2017).

48. Zapatero, V., van Leent, T., Arnon-Friedman, R., Liu, W.-Z., et al.: Advances in Device-Independent Quantum Key Distribution. NPJ Quantum Inform. 9, 10 (2023).
49. Li, W., Zhang, L., Tan, H., Lu, Y., et al.: High-Rate Quantum Key Distribution Exceeding 110 Mb/s. Nature Photon. 17, 416 (2023).
50. Sax, R., Boaron, A., Boso, G., Atzeni, S., et al.: High-Speed Integrated QKD System. Photon. Res. 11, 1007 (2023).
51. Grünenfelder, F., Boaron, A., Resta, G. V., Perrenoud, M., et al.: Fast Single Photon Detectors and Real-Time Key Distillation Enable High Secret-Key-Rate Quantum Key Distribution Systems. Nature Photon. 17, 422 (2023).
52. Mummadi, S., Rudra, B.: Practical Demonstration of Quantum Key Distri-bution Protocol with Error Correction Mechanism. Int. J. Theor. Phys. 62, 86 (2023).
53. Chen, Z., Wang, X., Yu, S., Li, Z., Guo, H.: Continuous-Mode Quantum Key Distribution with Digital Signal Processing. NPJ Quantum Inform. 9, 28 (2023).
54. Bennett, C. H., Brassard, G.: Quantum Cryptography: Public Key Distribution and Coin Tossing. Theor. Comp. Sci. 560, 7 (2014).
55. Pereira, M., Currás-Lorenzo, G., Navarrete, Á., Mizutani, A., et al.: Modified BB84 Quantum Key Distribution Protocol Robust to Source Imperfections. Phys. Rev. Res. 5, 023065 (2023).
56. Wang, D., Wang, H., Xu, H., Ji, Y.: Physical-Layer Encryption and Authentication Scheme Based on SKGD and 4D Hyper-Chaos. Opt. Expr. 7, 11829 (2023).
57. Arikan, E.: Channel Polarization: A Method for Constructing Capacity-Achieving Codes for Symmetric Binary-Input Memoryless Channels. IEEE Trans. Inform. Theory 55, 3051 (2009).
58. Sasoglu, E., Teletar, E., Arikan, E.: Polarization for Arbitrary Discrete Memoryless Channels. IEEE Inform. Theory Workshop (IEEE Press, New York, 2009) pp. 144–8.
59. Guo, J., Tang, B., Lai, T., Liang, X., et al.: The Implementation of Shannon-Limited Polar Codes-Based Information Reconciliation for Quantum Key Distribution. Quantum Sci. Technol. 8, 035011 (2023).
60. Cao, Z., Chen, X., Chai, G., Liang, K., Yuan, Y.: Rate-Adaptive Polar-Coding-Based Reconciliation for Continuous-Variable Quantum Key Distribution at Low Signal-to-Noise Ratio. Phys. Rev. Appl. 19, 044023 (2023).
61. Gao, H., Wang, K., Qu, D., Lin, Q., Xue, P.: Demonstration of a Photonic Router via Quantum Walks. New J. Phys. 25, 053011 (2023).
62. Mitra, A., Badhani, H., Ghosh, S.: Improvement in Quantum Communication using Quantum Switch. Phys. Scrip. 98, 045101 (2023).
63. Rochman, J., Xie, T., Bartholomew, J. G., Schwab, K. C., Faraon, A.: Micro-wave-to-Optical Transduction with Erbium Ions Coupled to Planar Photonic and Superconducting Resonators. Nature Commun. 14, 1153 (2023).
64. Sund, P. I., Lomonte, E., Paesani, S., Wang, Y., et al.: High-Speed Thin-Film Lithium Niobate Quantum Processor Driven by a Solid-State Quantum Emitter. Sci. Adv. 9, eadg7268 (2023).
65. Sosnicki, S., Mikolajczyk, M., Golestani, A., Karpinski, M.: Interface Between Picosecond and Nanosecond Quantum Light Pulses. Nature Photon. (2023).
66. Lago-Rivera, D., Rakonjac, J. V., Grandi, S., de Riedmatten, H.: Long-Distance Multiplexed Quantum Teleportation from a Telecom Photon to a Solid-State Qubit. Nature Commun. 14, 1889 (2023).
67. Giovannetti, V., Lloyd, S., Maccone, L.: Quantum Enhanced Measurements: Beating the Standard Quantum Limit. Science 306, 1330 (2004).
68. Macieszczak, K., Guta, M., Lesonovsky, I., Garrahan, J. P.: Dynamical Phase Transitions as a Resource for Quantum Enhanced Metrology. Phys. Rev. A 93, 022103 (2016).
69. Degen, C. L., Reinhard, F., Cappellaro, P.: Quantum Sensing. Rev. Mod. Phys. 89, 035002 (2017).

70. Johnson, J.: Thermal Agitation of Electricity in Conductors. Phys. Rev. 32, 97 (1928).

71. Nyquist, H.: Thermal Agitation of Electric Charge in Conductors. Phys. Rev. 32, 110 (1928).

72. Dicke, R. H.: The Measurement of Thermal Radiation at Microwave Fre-quencies. Rev. Sci. Instrum. 17, 268 (1946).

73. Handel, P. H., Chung, A. L.: Noise in Physical Systems and "1/f" Fluctuations. (AIP Publishing, New York, 1993).

74. Bongs, K., Bennett, S., Lohmann, A.: Quantum Sensors will Start a Revolution--if we Deploy them Right. Nature 617, 672 (2023).

75. Gorecki, W., Demkowicz-Dobrzanski, R., Wiseman, H. M., Berry, D. W.: π-Corrected Heisenberg Limit. Phys. Rev. Lett. 124, 030501 (2020).

76. Cimini, V., Polino, E., Belliardo, F., Hoch, F., et al.: Experimental Metrology Beyond the Standard Quantum Limit for a Wide Resources Range. NPJ Quantum Inform. 9, 20 (2023).

77. Shannon, C. S.: A Mathematical Theory of Communications. Bell Sys. Techn. J. 27, 379 (1948).

78. Landauer, R.: Irreversibility and Heat Generation in the Computing Process. IBM J. Res. Develop. 3 183 (1961).

79. Kurdzialek, S., Demkowicz-Dobrzanski, R.: Measurement Noise Susceptibility in Quantum Estimation. Phys. Rev. Lett. 130, 160802 (2023).

80. Pirandola, S., Bardhan, B. R., Gehring, T., Weedbrook, C., Lloyd, S.: Advances in Photonic Quantum Sensing. Nature Photonics 12, 724 (2018).

81. Burenkov, I. A., Jabir, M. V., Polyakov, S. V.: Practical Quantum-Enhanced Receivers for Classical Communication. AVS Quantum Sci. 3, 025301 (2021).

82. Chen, Y., Miao, Z., Yuan, H.: Cooperation Between Quantum Control and Noises in Quantum Metrology. Adv. Quantum Technol. 6, 2200165 (2023).

83. Gefen, T., Rotem, A., Retzker, A.: Overcoming Limits with Quantum Sensing. Nature Commun. 10, 4992 (2019).

84. Keyes, R. W., Landauer, R.: Minimal Energy Dissipation in Logic. IBM J. Res. Develop. 14, 152 (1970).

85. Keyes, R. W.: Fundamental Limits in Digital Information Processing. Proc. IEEE 69, 267 (1981).

86. Avery, J.: Information Theory and Evolution (World Scientific, New York, 2003).

87. Shiffer, M.: Shannon's Information is Not Entropy. Phys. Lett. A 154, 361 (1991).

88. Kish, L. B., Ferry, D. K.: Information Entropy and Thermal Entropy: Apples and Oranges. J. Comp. Electron. 17, 43 (2018).

89. Brillouin, L.: Science and Information Theory (New York, Academic Press, 1965).

90. Bate, R. T.: VLSI Electronics Microstructures, vol. 9 Einspruch N G, Ed (New York, Academic Press, 1982) 359–386.

91. Kish, L. B.: End of Moore's Law: Thermal (Noise) Death of Integration in Micro and Nano Electronics. Phys. Lett. A 305, 144 (2002).

92. Zhirnov, V. V., Cavin, R. K., Hutchby, J. A., Bourianoff, G. K.: Limits to Binary Logic Switch Scaling--A Gedanken Model. Proc. IEEE 91, 1934 (2003).

93. Gea-Banaloche, J. Kish, L. B.: Future Directions in Electronic Computing and Information Processing. IEEE Proc. 93, 1858 (2005).

94. Boltzmann, L.: Vorlesungen über Gastheorie, vol. 1. (Leipzig, Barth, 1896) Sect. 2.6.

95. Planck, M.: The Theory of Heat Radiation. Tr. Massius, M. (Philadelphia, Blakiston, 1914) p. 119.

96. Ferry, D. K., Porod, W.: Dissipation and Irreversibility in Computing. Mat. Res. Express, in press.

97. Chang, J., Gao, J., Zadeh, I. E., Elshaari, A. W., Zwiller, V.: Nanowire-Based Integrated Photonics for Quantum Information and Quantum Sensing. Nanophoton. 12, 339 (2023).
98. Aharonovich, I., Englund, D., Toth, M.: Solid-State Single-Photon Emitters. Nature Photon. 10, 631 (2016).
99. Ilyakov, I., Ponomaryov, A., Reig, D. S., Murphy C., et al.: Ultrafast Tunable Terahertz-to-Visible Light Conversion through Thermal Radiation from Graphene Metamaterials. Nano Lett. 23, 3872 (2023).
100. Ge W., Jacobs, K., Zubairy, M. S.: The Power of Nonclassical States to Amplify the Precision of Macroscopic Classical Metrology. NPJ Quantum In-form. 9, 5 (2023).
101. Pucher, S., Liedl, C., Jin, S., Rauschenbeutel, A., Schneeweiss, P.: Atomic Spin-Controlled Non-Classical Raman Amplification of Fibre-Guided Light. Nature Photon. 16, 380 (2022).
102. Sinha, K., Goldschmidt, E. A.: An Atomic Spin on Amplification of Light. Nature Photon. 16, 337 (2022).
103. Das, S., Zhang, J., Marina, S., Suter, D., Caruso, F.: Quantum Pattern Recognition on Real Quantum Processing Units. Quantum Mach. Intell. 5, 16 (2023).
104. Peng, Y., Qin, S., Zhang, S., Zhao, Y.: Optical Fiber Quantum Temperature Sensing Based upon Single Photon Interferometer. Opt. Lasers Engr. 167, 107611 (2023).
105. Weinbub, J., Ferry, D. K.: Recent Advances in Wigner Function Approaches. Appl. Phys. Rev. 5, 041104 (2018).
106. Ferry, D. K., Nedjalkov, M.: The Wigner Function in Science and Technology. (Bristol, IOP Publishing, 2018).
107. Wigner, E. P.: On the Quantum Correction to Thermodynamic Equilibrium. Phys. Rev. 40, 749 (1932).
108. Birrittella, R. J., Alsing, A. M., Gerry, C. C.: The Parity Operator: Applications in Quantum Metrology. AVS Quantum Sci. 3, 014701 (2021).
109. Podoshvedov, M. S., Podoshvedov, S. A.: Family of CV States of Definite Parity and Their Metrological Power. Laser Phys. Lett. 20, 045202 (2023).
110. Assouly, R., Dassonneville, R., Peronnin, T., Bienfait, A., Huard, B.: Quantum Advantage in Microwave Quantum Radar. Nature Phys. 19, in press (2023).
111. Girard, J.-P., Lake, R. E., Liu, W., Kokkoniemi, R., et al.: Cryogenic Sensor Enabling Broad-Band and Traceable Power Measurements. Rev. Sci. In-strum. 94, 054710 (2023).
112. Huang, X.-J., Han, P.-R., Ning, W., Yang, S.-B., et al.: Entanglement-Interference Complementarity and Experimental Verification in a Super-conducting Circuit. NPJ Quantum Inform. 9, 43 (2023).
113. Toida, H., Sakai, K., Teshima, T. F., Hori, M., et al.: Magnetometry of Neurons using a Superconducting Qubit. Commun. Phys. 6, 19 (2023).
114. Boss, J. M., Cujia, K. S., Zopes, J., Degen, C. L.: Quantum Sensing with Arbi-trary Frequency Resolution. Science 356, 837 (2017).
115. Zhou, H., Choi, J., Choi, S., Landig, R., et al.: Quantum Metrology with Strongly Interacting Spin Systems. Phys. Rev. X 10, 031003 (2020).
116. Yoon, J., Kim, K., Na, Y., Lee, D.: Characterization and Correction of the Pulse Width Effects on Quantum Sensing Experiments Using Solid-State Spin Qubits. Curr. Appl. Phys. 50, 140 (2023).
117. Vorobyov, V., Zaiser, S., Abt, N., Meinel, J., et al.: Quantum Fourier Transform for Nanoscale Quantum Sensing. NPJ Quantum Inform. 7, 124 (2021).
118. Yu, C.-J., von Kugelgen, S., Laorenza, D. W., Freedman, D. E.: A Molecular Approach to Quantum Sensing. ACS Cent. Sci. 7, 712 (2021).

The Josephson-Based Qubit

3

As was discussed in Sect. 1.3, many (if not most) of the commercial efforts in quantum computing (and the other areas) lie with superconductivity and the use of Josephson junctions for qubits. This brings us to the use of quantum physics to actually create quantum bits, and also quantum circuits. The physics of superconductivity, and the Josephson junction (JJ), is rather deep. However, an overview of superconductivity and the JJ will be presented first in this chapter in order to acquaint the reader with the basis for this technology in the quantum world. Then, the discussion will move to the superconducting quantum interference device (SQUID) and qubits.

3.1 Superconductivity

The first person to liquify helium is generally considered to be Heike Kammerlingh Onnes (later he would drop his first name in favor of the shorter version), a Dutch physicist, in 1908 [1]. Helium becomes liquified at temperature below 4.15 K ($-452.2F$). Onnes soon was busy investigating the low-temperature properties of any material he could put into his cryostat in his lab in Leiden. Three years later, he was investigating mercury, and as he cooled the mercury below 4.2 K, a strange phenomenon occurred, in which the resistance of the mercury abruptly dropped to an exceedingly low value (measured to be less than 10^{-6} Ω at 3 K). Today, Onnes' discovery is known as *superconductivity*, and is observed in over one-fourth of the elements of the periodic table, and in numerous compounds and alloys. The general behavior of the material is depicted in Fig. 3.1. Normally, one would expect that the resistance of a metal would decrease with temperature (the blue dashed line in the figure). Instead, the resistance drops abruptly at a temperature

© The Author(s), under exclusive license to Springer Nature Switzerland AG 2025   63
D. K. Ferry, *Quantum Information in the Nanoelectronic World*,
Synthesis Lectures on Engineering, Science, and Technology,
https://doi.org/10.1007/978-3-031-62925-9_3

Fig. 3.1 The temperature variation of the resistance in a superconducting material. The dashed curve illustrates the expected behavior, while the red curve shows the abrupt transition to the superconducting state

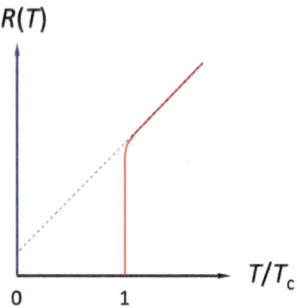

known as the *transition temperature* or *critical temperature* (T_c), and below this temperature the material is superconducting and the resistance is essentially zero. The transition to the superconducting state is considered to be a phase transition.

The transition temperature of the material is the easiest parameter to measure, as may be seen from Fig. 3.1. There is a slight rounding of the resistance curve just above T_c, and the enhanced conduction at this point is sometimes referred to as *paraconductivity*. This enhanced conductivity is attributed to thermal fluctuations that generally exist at any phase transition, and one can think of them as small regions, in which superconducting behavior is beginning to occur. These would be small domains of the overall sample. Superconductivity of the entire sample begins at T_c. The interactions which lead to superconductivity tend to be in poor metals, while good metals (such as gold and silver) do not generally exhibit superconducting behavior. There are a number of compounds which exhibit superconductivity well above the temperature at which it is seen in Hg.

A sufficiently large magnetic field applied to the material will destroy the superconductivity. There is a critical value of the magnetic field, different for different materials, at which the superconductivity is destroyed and the material reverts to the normal state. This critical field H_c is a function of temperature and varies normally with the relationship

$$H_c = H_{c0}\left[1 - \left(\frac{T}{T_c}\right)^2\right], T < T_c \tag{3.1}$$

The subscript 0 has been added on the right-hand side to indicate the value at absolute zero. The behavior of (3.1) is that followed by (what are called normal or type 1) superconductors. There is another variation which occurs in what are called type-2 superconductors, in which the magnetic field itself begins to affect H_c. This is reflected in the magnetization M of these type-2 materials, in which M increases linearly with applied magnetic field up to a lower transition field H_{c1}, then decreases with further field increases until it disappears at H_{c2}. This type-II behavior is important because most of the materials that are used to make superconducting magnets are, in fact, type-II materials.

The general understanding of superconductivity, at least as this occurs in most materials was developed in 1957 by Bardeen, Cooper, and Schrieffer [2]. The basic interaction is

Fig. 3.2 Superconductivity opens a gap $E_G = 2\Delta$ in the energy spectrum of the electrons around the Fermi energy E_F

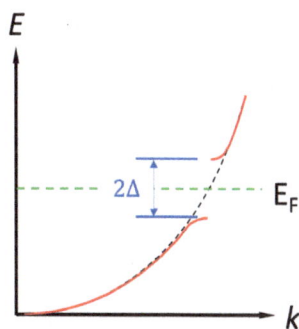

that an attractive force between a pair of electrons, that have opposite spins, can lead to a lower energy state at low temperatures. The actual interaction is from the atom lattice itself, in that the interaction one of the electrons creates a local deformation which then attracts the second electron, provided it has opposite spin. The paired electrons form a new particle state termed a Cooper pair [3]. In order for the pair to form, the interaction has to lead to a lower energy state, and this results in a gap opening in the energy band between paired electrons and unpaired electrons. This is shown in Fig. 3.2. The normal relationship between the energy E and the momentum wave number k is shown by the black dashed line. The actual relationship is shown by the red curves, and this indicates the gap opening around the Fermi energy E_F. The size of this gap is denoted as 2Δ, the 2 for the two electrons and then Δ is the energy per electron in the superconducting transition. As with the magnetic field, the gap varies with the temperature and, in this case, the form is.

$$E_G = 2\Delta = E_{G0}\sqrt{1 - \left(\frac{T}{T_c}\right)},\qquad(3.2)$$

where the subscript 0 on the right-hand side indicates the value at absolute zero temperature. Obviously, this energy gap has a relationship to the transition temperature, but this relationship varies with the material. Over a large range of materials, the connection varies somewhat as $E_{G0}/k_B T_c \sim 3.5 - 4.5$. At 4 K, the thermal energy is $k_B T \sim 5.5 \times 10^{-23} J = 0.345\text{meV}$. So the zero temperature gap energy is of the order of 1 or a few meV.

3.2 The Josephson Junction

The Josephson junction is formed from two superconducting materials that are separated by an insulating layer [4, 5]. The insulating layer needs to be thin enough that a sufficient tunnel current can flow through it. Normally, this current is composed of single particles

Fig. 3.3 The current–voltage characteristic for the Josephson junction. At zero temperature, the single-particle current would be the dashed red curve, but at non-zero temperature, the blue curve results. When the insulator is sufficiently thin, the Cooper pairs can tunnel, giving the d.c. Josephson effect seen as the current in the green arrow

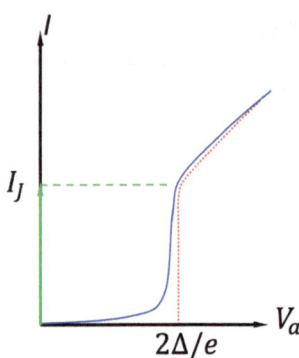

and not Cooper pairs. The principle of operation is described in Fig. 3.3. The dashed red curve is the current that would flow at zero temperature; that is, there is no current until a sufficient voltage is applied to break up the Cooper pairs ($eV_a \geq 2\Delta = E_G$). The existence of the gap in the spectrum, as shown in Fig. 3.2, means that the states on one side of the insulator must be raised sufficiently high in energy that they overlap the states above the gap on the other side of the insulator. There are no states available in the gap, so the occupied states below the gap have to be raised in energy so that they overlap the available unoccupied states on the other side of the junction. Tunneling. is usually an energy conserving process, so that the energy of the tunneling particles must be larger than the gap. This energy is sufficient to break up the Cooper pairs. However, at temperatures above zero, but still below the transition temperature of the superconductors, there are unpaired electrons which can tunnel, and this leads to the blue curve in Fig. 3.3. So far, this is just a normal tunnel junction, albeit one with the superconducting gap present.

New behavior, characteristic of the Josephson junction, begins when the insulating layer becomes sufficiently thin that the wave functions of the Cooper pairs can extend through the insulator. This allows the development of phase coherence between the wave functions on the two sides of the junction, and this allows the Cooper pairs to begin to tunnel through the insulator. This is shown by the green current curve at $V_a = 0$. Since the Cooper pairs are superconducting, they offer no resistance and can tunnel at basically zero voltage. This is known as the d.c. Josephson effect. The maximum, or critical, current here is I_J, and when this is exceeded, the characteristic jumps to the single-particle curve as shown by the dashed green line, and then follows the blue line at higher voltages. The critical current I_J depends upon properties of the junction structure and materials. The actual supercurrent (green arrow) that flows through the junction is related to the phase difference between the two superconducting materials on either side of the junction through the relationship.

$$I_S = I_J sin[\varphi(t)]. \tag{3.3}$$

The time-dependence of the phase in (3.3) is to indicate that there is also an a.c. Josephson effect.

The a.c. Josephson effect arises from the fact that an applied voltage will cause a change in the phase difference between the two superconductors. This change in phase is closely related to Faraday's law for magnetic induction, as the phase change can be expressed as.

$$V_a = \frac{d\Phi}{dt}, \ \Phi = \frac{h}{2e}\frac{\varphi}{2\pi}. \tag{3.4}$$

This relationship tells us that the phase oscillates in an applied voltage, and this can lead to photon emission with a frequency.

$$f = \frac{2eV_a}{h}. \tag{3.5}$$

Of course, there is an inverse process in which photons can be absorbed and lead to an induced voltage, simply by rearranging (3.5). When microwave absorption occurs it leads to steps in the blue curve of Fig. 3.3 in the region below the energy gap. These steps can arise from either Cooper pair absorption or from single particle photon-assisted tunneling. In the case of superconducting pairs, steps in the current–voltage curve arise at voltages given by.

$$V_a = n\hbar\omega/2e, \tag{3.6}$$

and are often called Shapiro steps [6]. Photon-assisted tunneling of single particles can also occur at voltages of twice the value in (3.6) (no factor of 2 for the Cooper pairs).

The connection to Faraday's law, and the fact that the quantity $\Phi_0 = h/2e$ is the quantum of flux arising from the quantization of the magnetic field (the factor of 2 arises from considering the Cooper pairs). This connections suggests that circuits with Josephson junctions will also be good detectors for magnetic fields. Later, interest will focus on a quantity called the Josephson energy, that is related to the critical current in Cooper pair tunneling by

$$E_J = \frac{hI_J}{4\pi e}. \tag{3.7}$$

As noted above, the Josephson junction can be a microwave source (and detector). When acting as a source, there is an inverse effect in which a so-called Riedel peak appears in the response of the junction [7, 8]. The Josephson frequency is given by (3.5), and the Riedel peak is when this frequency also satisfies the relation $f = 4\Delta/h$, that is at a frequency corresponding to the energy gap.

When the Josephson junction device is physically small, in the sense that the current carrying area is quite small, then another energy scale enters the picture. This is from the Coulomb blockade [9, 10]. While an old idea, it is today quite important in the work of nanoscale devices [11]. The important factor is the capacitance of the tunnel junction, which is given by the ratio of the area of the junction to the thickness of the dielectric material ($C = \epsilon A/d$, where ϵ and d are the dielectric function and the thickness of the insulator, and A is the junction area). When an electron tunnels through the insulator, its energy changes by the amount.

$$\Delta E = \frac{e^2}{2C} \tag{3.8}$$

In the normal world, this energy is much larger than the thermal energy $3k_B T/2$, and tunneling can be driven merely by thermal fluctuations. In the nano-world, this is no longer true. If (3.8) is smaller than the thermal energy, tunneling cannot occur; this is the blockade world. Thus, in a blockaded device with small capacitance, a voltage of $\pm e/C2$ must be applied to provide sufficient energy to the tunneling electron. It turns out that when superconducting qubits are created using Josephson junctions, then the dependence upon the two energies (3.7) and (3.8) and their relative sizes will be enhanced by adding parts that really depend upon (3.8), the so-called single-electron circuit. This will be discussed in the Sect. 2.4, before the physics of qubits is considered. In the modeling of the single-electron tunneling, it is found that the Riedel peak and the gap for the quasi-particle-pair interference are shifted to $2\Delta/e - e/2C$, while the quasi-particle[1] tunneling gap shifts to $2\Delta/e + e/2C$ [12]. Hence, there is a hybridization between the superconducting gap and the Coulomb gap.

3.3   The SQUID

A superconducting quantum interference device (SQUID) is a special circuit in which one or more Josephson junctions are embedded. It was originally developed to serve as very sensitive magnetometer to measure magnetic fields (such as in Magnetic Resonance Imaging, MRI) commonly used today [13]. The most common form of the SQUID is a wire loop with two Josephson junctions embedded in the loop, much as in Fig. 3.4, where the SQUID is shown in red. When a bias current is applied to the ring, this current divides between the two possible paths, so that $I_b/2$ flows through each arm of the circuit. Assuming the two junctions are equal, each has a critical current of I_J, governed by (3.3). Now, if a small *external* magnetic field is applied, a screening current I_s will be induced (a byproduct of Faraday's Law), that induces a change in the phase at each of the junctions. This screening current flows around the red loop in the figure, so that one junction has a current $I_b/2 + I_s$ and the other has a current $I_b/2 - I_s$. Normally, there is no voltage at the output (indicated in green in the figure), since the loop is superconducting. But, as soon

---

[1] The term quasi-particle is used to denote the fact that these are not real particles, but are new "things" created from combinations of particles. In this case the Cooper pair is such a quasi-particle. Other versions will appear later.

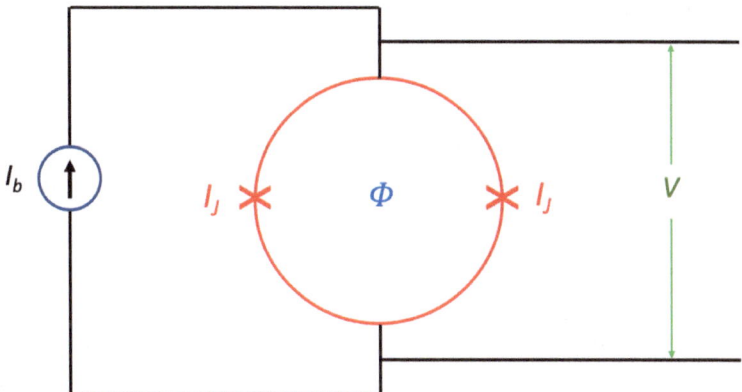

Fig. 3.4 Basic circuit of a SQUID under current bias. The red section is the actual SQUID–a loop with two Josephson junctions embedded in it. When a bias current I_b is passed through it, a voltage V is developed, which can have complicated behavior, discussed in the text

as one of the two currents exceeds the critical current I_J in that junction, a voltage will be developed since that junction is now resistive (see Fig. 3.3).

Increasing the flux so that it exceeds $\Phi_0/2$ leads to new behavior. This is because the magnetic flux contained within the loop is quantized and must be an integer multiple of Φ_0. Now, the current through the junction must change so that it reflects the *difference* in the flux inside the loop (Φ_0) and the flux outside the loop $> \Phi_0/2$. This causes the junction current to decrease as the external flux is increased until the latter reachs Φ_0. When this occurs, the fluxes inside the loop and outside the loop are equal, and the effective field seen by the junction vanishes, and the current is zero. Thus, as the magnetic field is increased in amplitude, the junction current is going to oscillate with a period of Φ_0, with peak current at $(n + 1/2)\Phi_0$ and a minimum current at $n\Phi_0$.

If the bias current I_b is larger than the critical current of the junctions, then this behavior is not seen as the device works only in a resistive mode. Nevertheless, the output voltage will show oscillations that are periodic in the magnetic field, with a period of Φ_0. Hence this device can act as a flux to voltage convertor, and sensitive measurements can be made on the actual magnetic field. Obviously, two SQUIDs can be connected in a manner to measure the derivative of the magnetic field, four SQUIDS for the second derivative and so on. Derivatives are often used for more sensitive measurements of magnetic field variations. If microwave fields are used, then the SQUID can work effectively with only a single Josephson junction.

3.4   Single-Electronics

While the SQUID is an excellent sensor, especially for magnetic fields, its use as a qubit can be more tenable by extending the circuit to contain single-electron sensitive circuits that depend upon the Coulomb blockade of (3.8). As was remarked in Sect. 3.2, when an electron moves across a capacitor (from one plate to the other, often by tunneling), the change in electrostatic potential may be expressed as.

$$\Delta V = \frac{e}{C}, \tag{3.9}$$

where e is the charge that transits/tunnels through the capacitor C. Normally, in a capacitor, the two plates are metal conductors, that have a charge $+Q$ on one conductor and a charge $-Q$ on the other, when the voltage V is applied. When the capacitor is charged in such a manner, the electrostatic energy stored on the two conductors and is given by.

$$E = \frac{Q^2}{2C}. \tag{3.10}$$

This is, of course, just (3.8) when one considers a single electron charge. Just as discussed above, when the charging energy given by this latter equation due to a single electron, $e^2/2C$, becomes larger than the thermal energy, $k_B T$, the transfer of a single electron between the two conductors becomes thermally blocked–the Coulomb blockade regime. The blockage remains until the charging energy is overcome by a sufficiently large applied voltage. While the ideas behind the Coulomb blockade are old [9, 10], they were brought back to the forefront of science by later experiments on small metallic capacitors [11, 14]. Typical current–voltage curves are shown in Fig. 3.5 for tunneling between the small tungsten tip of a scanning tunneling microscope (STM) and a stainless steel metallic surface [15]. Here, the capacitance arises from the air gap between the tip and the surface, and the system is immersed in pumped liquid helium at~1 K. The straight lines are guides to the eye, but indicate roughly the blockade voltage needed to tunnel. Measurements on the system give values for the effective capacitance C and the tunneling resistance R, so that one can estimate the expected current to vary with the applied voltage as [15]

$$I = \frac{2C}{\pi e R} V^2, \tag{3.11}$$

and this is the thin quadratic curve through the data points near the origin. This shows good agreement with the expected current.

A Single-Electron Circuit. While the single capacitor is interesting to describe blockade, real circuits to utilize the effect normally include at least a pair of capacitors. The concept of a single-electron device (or transistor) uses (at least) two capacitors connected in series. Between these two capacitors, there exists a region often referred to as a "dot" or "box". In semiconductors, one can usually totally deplete this region of charge, so that dot

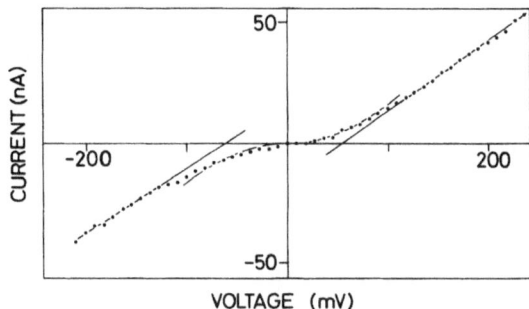

Fig. 3.5 *I-V* characteristic for tunneling between a tungsten STM tip and a stainless steel surface. The quadratic curve near the origin is the expected current from (3.11). The straight lines are merely guides to the eye. Reprinted with permission from [15], copyright 1988 by the American Physical Society

is commonly used. With metals, this is not the case, so that box is normally used, particularly in connection with superconducting qubits. In this latter case, the box may contain more than a thousand electrons, although it is the macroscopic charge non-neutrality that creates the macroscopic quantum state which can be changed by the transfer of a single electron, or a Cooper pair.

Consider the circuit of Fig. 3.6, in which two capacitors are shown in series, with a quantum dot/box region between them. The two capacitors have thin insulators that allows tunneling, although they are small enough that this tunneling is blockaded. Typically, each of these two tunneling capacitors can be considered as a parallel combination of the tunneling resistance R and the physical capacitance C. (The resistance part describes current flow when the device is in the tunneling state, but remains large, $> h/2e^2$.) This resistance is not important to the development following below, but should not be forgotten. The interest is in the charge that can accumulate in/on the quantum box (shown in red in the figure) in a sequential tunneling approach. That is, it is considered that charge may tunnel through one, or the other, of the two capacitors. Tunneling through just one of the capacitors will change the amount of charge in the box.

The charge residing on the electrodes of the individual capacitors is given by Eq. (3.9) (but with the total charge Q rather than that of a single electron), so that the two capacitors yield.

$$Q_1 = C_1 V_1$$
$$Q_2 = C_2 V_2. \tag{3.12}$$

The net charge Q_{box} in the box is the difference of these two charges, and arises from tunneling through one or the other of the two capacitors. This charge may be expressed as.

Fig. 3.6 A single-electron circuit containing two capacitors and a quantum box located between them. The voltage source on the left provides bias that will be used to describe the operation of this circuit

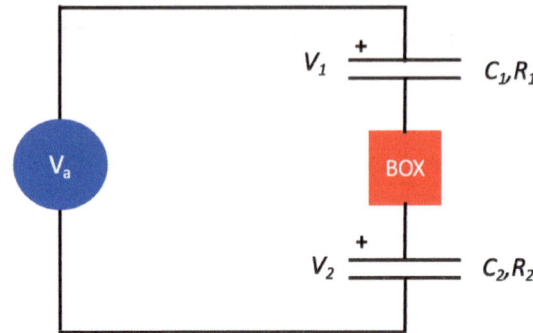

$$Q_{box} = -ne = Q_2 - Q_1. \tag{3.13}$$

Here, n is the net number of electrons that have accumulated in the box. The sign convention is chosen so that an increase in either n_1 or n_2, the number of electrons on the two capacitors, corresponds to increasing the respective charge Q_1 or Q_2, in Eq. (3.12). The sum of the two voltages on the two capacitors gives the applied voltage, V_a, and the two voltage drops may be found as (from normal circuit theory).

$$\begin{aligned} V_1 &= \tfrac{1}{C_T}(C_2 V_a + ne) \\ V_2 &= \tfrac{1}{C_T}(C_1 V_a - ne) \end{aligned}, \tag{3.14}$$

where $C_T = C_1 + C_2$ is the total capacitance in the circuit. Now the electrostatic energy can be found using (3.10) for the two capacitors to be.

$$E = \frac{Q_1^2}{2C_1} + \frac{Q_2^2}{2C_2} = \frac{1}{2C_T}\left(C_1 C_2 V_a^2 + Q_{box}^2\right) \tag{3.15}$$

However, the result (3.15) is not all of the energy in the circuit. This is because one must add to this result the work that is done by the voltage source in moving charge to the capacitors. This involves the various tunneling currents, and the amount of charge in each capacitor. Each capacitor thus provides its own contribution, and these may be expressed as [16]

$$\begin{aligned} W_a(n_1) &= -n_1 e V_a C_2 / C_T \\ W_a(n_2) &= -n_2 e V_a C_1 / C_T \end{aligned}. \tag{3.16}$$

Using (3.16) with (3.15), the total energy in the system of Fig. 3.6 is now given as.

$$E = \frac{1}{2C_T}\left(C_1 C_2 V_a^2 + Q_{box}^2\right) + \frac{eV_a}{C_T}(C_1 n_2 + C_2 n_1) \tag{3.17}$$

The condition that is necessary for Coulomb blockade to exist arises from considering the *change* in this electrostatic energy when a particle tunnels through either capacitor. At zero temperature, thermodynamics (and energy considerations) requires that the system always evolves from a state of higher energy to one of lower energy. Tunneling transitions that would cause the system to transition to a state of higher energy cannot occur at zero temperature, unless driven by even higher voltages. At the other end of the scale, when the temperature is sufficiently high, the thermal fluctuations wash out the Coulomb blockade, and the capacitors are merely leaky. To examine the blockade, we first assume that the charge on C_2 changes by the addition or subtraction of a single electron. This leads to the energy change.

$$\Delta E_2^{\pm} = E(n_1, n_2) - E(n_1, n_2 \pm 1) = \frac{e}{C_T}\left[-\frac{e}{2} \pm (ne - V_a C_1)\right]. \tag{3.18}$$

Similarly, when the charge on C_1 changes by the addition or subtraction of a single electron, the energy change is.

$$\Delta E_1^{\pm} = E(n_1, n_2) - E(n_1 \pm 1, n_2) = \frac{e}{C_T}\left[-\frac{e}{2} \mp (ne + V_a C_2)\right]. \tag{3.19}$$

For all possible transitions of charge into or out of the box, the leading term involving the Coulomb energy of the island causes ΔE to be negative until the magnitude of V_a exceeds a threshold value. When the capacitances are equal ($C_1 = C_2 = C$), the result is $|V_a| > e/2C$, which is the simple result already discussed below (3.8). This potential range for which Coulomb blockade exists, is a result of the additional Coulomb energy, $e^2/2C_T$, that must be expended by an electron when it tunnels into or out of the box. The effect on the current voltage characteristics is a region of very low conductance around the origin, as shown in Fig. 3.5 for the STM tunneling experiment. But, this simple circuit can be augmented to provide greater control of the charge. The structure is often called a single-electron transistor (SET).

Adding a Gate Circuit. The desired Coulomb blockade is easily obtained with the single-electron circuit of Fig. 3.6, but this circuit is both basic and somewhat difficult to control, if we are to use it in a qubit. Fortunately, it is easy to add another bias voltage to gain greater control of the actual charge in the box. Consider the circuit shown in Fig. 3.7. In this circuit, a separate voltage source, V_g, is coupled to the box through the capacitor, C_g. This latter capacitor is a non-tunneling capacitor, as this additional circuitry is merely to better control the charge in the box. This means that an additional charge

$$Q_g = C_g(V_g - V_2) \tag{3.20}$$

will appear in the box. The net charge in the box modifies (3.13) so that it now becomes

$$Q_{box} = -ne = Q_2 - Q_1 - Q_g. \tag{3.21}$$

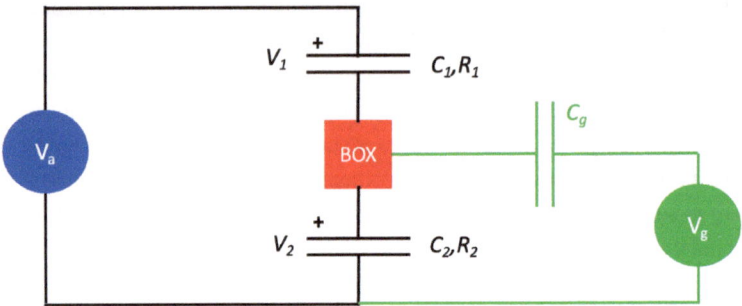

Fig. 3.7 Adding the additional circuitry (in green) with bias V_g and capacitance C_g gives better control of the charge in the box (red)

Needless to say, all of the various equations above will be modified accordingly. The new forms of the voltages across the first two capacitors also are modified to include the effect of the gate and its capacitance. Thus, Eqs. (3.14) will now become.

$$V_1 = \frac{1}{C_T}\big[(C_2 + C_g)V_a - C_g V_g + ne\big]$$
$$V_2 = \frac{1}{C_T}(C_1 V_a + C_g V_g - ne) \quad ,$$

(3.22)

with the total capacitance now being $C_T = C_1 + C_2 + C_g$. Similarly, the total energy on the three capacitors changes (3.15) to.

$$E = \frac{1}{2C_T}\Big[C_1 C_2 V_a^2 + C_g C_1 \big(V_g - V_a\big)^2 + C_g C_2 V_g^2 + Q_{box}^2\Big].$$

(3.23)

The work performed by the two voltage sources arising from the current through the two tunneling junctions 1 and 2 must now include both the work done by the gate voltage and the additional charge flowing into the box from the gate capacitor. Equations (3.16) now become.

$$W_a(n_1) = -n_1 e\big[V_a C_2 - C_g(V_g - V_a)\big]/C_T$$
$$W_a(n_2) = -n_2 e\big[V_a C_1 + V_g C_g\big]/C_T \quad .$$

(3.24)

Finally, the change in energies (3.18) and (3.19) for charging, or discharging, the box, now take the form.

$$\Delta E_2^{\pm} = E(n_1, n_2) - E(n_1, n_2 \pm 1)$$
$$= \frac{e}{C_T}\Big[-\frac{e}{2} \pm \big(ne - V_a C_1 - V_g C_g\big)\Big],$$

(3.25)

and

$$\Delta E_1^{\pm} = E(n_1, n_2) - E(n_1 \pm 1, n_2)$$

$$= \frac{e}{C_T} \left[-\frac{e}{2} \mp \left(ne - V_g C_g + V_a C_2 + V_a C_g \right) \right]. \tag{3.26}$$

The addition of the gate bias allows one to change the charge in the box in a manner that shifts the region in which Coulomb blockade is observed. This yields a stable region of Coulomb blockade that may exist even for $n \neq 0$. The condition for tunneling at low temperature remains $\Delta E_{1,2} > 0$, and still results in the system going to a state of lower energy when tunneling. However, now the equations for tunneling in various directions are modified to require.

$$-\frac{e}{2} \pm \left(ne - V_a C_1 - V_g C_g \right) > 0$$
$$-\frac{e}{2} \pm \left(ne - V_a C_1 - V_g C_g \right) > 0. \tag{3.27}$$

There are now four equations in (3.26), a selected value for n generates a stability plot in the (V_a, V_g) plane. In this plot, shown in Fig. 3.8, stable regions corresponding to each n that can exist, and no tunneling may occur in these regions. This diagram is for a choice of capacitors $C_g = C_2 = C$, $C_1 = 2C$. The various lines represent the boundaries for the onset of tunneling and are given by the four Eqs. (3.26) for different values of n. The trapezoidal shaded areas (shown in light blue) correspond to regions where tunneling may not exist. This means that the Coulomb blockade islands are much larger, and are much easier to control. Nevertheless, there are values of the two voltages for which tunneling easily occurs (where the lines cross one another), and the charge in the box can easily be changed. This is important for changing the qubit state, as will be seen in the next few sections. Indeed, the plot in Fig. 3.8 tells us that there are sets of voltages for which a single charge can be added or removed from the box and this fact is utilized in the qubits.

3.5 Superconducting Qubits

Moving from a single Josephson junction, or even a pair of junctions in a SQUID, toward a qubit seems to be a natural step, as for example, with experiments that couple a microwave cavity to a superconducting qubit device [17]. The interest in superconducting qubits goes back further. There are two basic types of superconducting circuits. These describe microscopic and macroscopic quantum objects [18]. At one end of the scale is the microscopic object based upon an internal charge or spin or two natural levels as in an atom. At the other end are macroscopic objects based on collective degrees of freedom like the persistent current in a superconducting ring or SQUID [19]. This can even be excess charge on a metallic superconducting island added to a SQUID [20, 21]. Although one might think that the macroscopic form will be physically larger, this is an unfounded viewpoint as all of these devices can be fabricated at the nanoscale. The

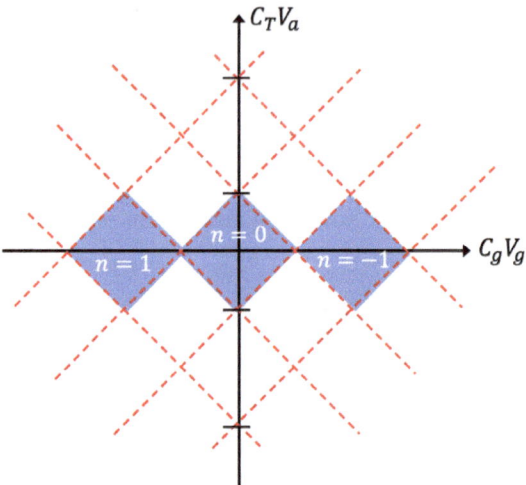

Fig. 3.8 The charging diagram for the circuit of Fig. 3.7 in the situation for which $C_g = C_2 = C$, $C_1 = 2C$. The various dashed (red) lines arise from Eqs. (3.26) for different values of charge in the box. The blue boxes are regions where the Coulomb blockade forbids tunneling to occur. Where the lines cross, the charge in the box can be changed by a single electron. The quantity CV is a charge, so the diagram is often called a *charging diagram*. For example, changing V_g, as one moves from the left side to the right side, successively removes a single charge from the box as each crossing of the red lines is passed

macroscopic object, or qubit, arises from the fact that when superconducting components become electrical circuits, then these circuits may become quantum objects in their own right. As we will see below, all of these approaches arise from the discussions above. The pursuit of superconducting qubits may utilize all of the above, be it a Cooper-pair box charge qubit, a persistent current flux qubit, or a hybrid charge-flux qubit [18]. As a group, this was seen in the first approach adopted by IBM for its initial quantum computers [22].

In the subsequent years, there have been many advances in the art of superconducting qubits. Fabrication has found new approaches which can involve flip-chip processing [23], reducing junction area fluctuations [24], or enhancing the coherence of the qubits with an applied electric field [25]. An important part of this is improving the distance over which entanglement can be maintained [26]. These improvements help the development beyond early (disputed) claims of achieving quantum supremacy (reaching the point at which quantum computing is demonstrably better than classical computing) [27].

As will be seen, there are two critical energies in the circuits utilizing both a small size and the SQUID. These are more evident when a quantum box is added to the SQUID, when the box provides single electron control, as discussed above. These are the Josephson energy (3.7) and the single-electron charging energy (3.8). These are central to the charge qubit and the flux qubit, both mentioned above. More importantly, they tend to

Fig. 3.9 A charge qubit as used in the work of ref. [20]. The basic qubit is the SQUID circuit shown in red, which is modified by the charge box, shown in blue. The voltages and capacitors, along with the flux, allow thorough control of the operation of the circuit

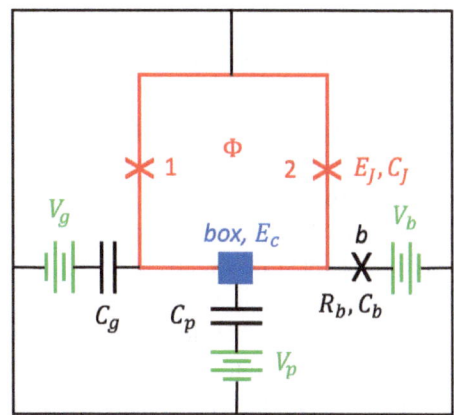

be central to all qubits in superconducting circuits. How these two energies are related suggests how the qubit will likely function. In general, if one thinks about the quantum levels arising from the charge in the box (see Fig. 3.9 below), one can view the box as an artificial atom, while the SQUID plays the role of a quantized resonator, so that the qubit becomes analogous to the Jaynes-Cummings model of Sect. 2.3. In the charge qubit, the single-electron charging energy dominates the circuit, while in the flux qubit the reverse is true. There is also a hybrid situation when the two energies are comparable in size. Finally, there is a more modern configuration known as a *transmon* qubit, which is used quite often today. These forms of qubit will be examined more fully in the following paragraphs.

The Charge Qubit. The charge qubit, like many other forms, is generated by inserting a box, or quantum dot, into the SQUID circuit, as shown in Fig. 3.9. Here, the primary SQUID (with the Josephson junctions labeled as 1 and 2) is shown in red, with a current flowing within this loop, and likely a phase difference between the two Josephson junction phases. The new item is the box, indicated in blue which has a single-electron charging energy of E_C, although it will primarily be used with Cooper pairs. This box could be a quantum dot of one kind or another, but is often formed by a pair of small Josephson junctions that isolate the region between them. Then single electron tunneling (or Cooper pair tunneling) governs the excitation of the box itself. When a single Cooper pair sits in this box [20, 28], it becomes a unique solid-state system in its own right. While there may be a relatively large number of electrons actually in the box, they should all be formed into Cooper pairs in the superconducting state and these then condense into a macroscopic quantum state. This macrostate is separated from the normal electrons by the superconducting gap 2Δ. The only low-energy excitations that are connected with the box arise from Cooper pair tunneling when 2Δ is larger than the charging energy E_C. In the charge qubit, the charging energy is larger than the gap energy, so that the thermal fluctuations are suppressed. As a result, the system becomes a simple two level system. where the two energy states differ by the occupation energy of one Cooper pair. These two states, that differ by one Cooper pair, are denoted as $|0\rangle$ and $|2\rangle$ (differing by the two electrons that form the Cooper pair). Thus, this charge qubit is easily related to the two-level system discussed in Sect. 2.2, and the two states can be adjusted so that they come into resonance with each other when the various voltages are suitably adjusted, as in Sect. 3.4 above and in Fig. 3.8.

This separation in the two energy levels can be controlled by controlling the number of electrons (Cooper pairs) in the box with an additional gate voltage, denoted as V_g (or V_P), coupled to the box through the capacitor $C_g(C_p)$, shown in Fig. 3.9. There is an additional voltage shown in the circuit: V_b, which is coupled through an additional Josephson junction (labeled b). Both of these additional bias voltages are coupled directly to the SQUID circuit.

In this circuit, the electrostatic energy of the box is controlled by the gate voltage V_g, so that this energy is set to be $E_C(n - q_t^2/e)$, where is the number of electrons in the box, and q_t is the gate induced charge. When this latter is just one electron charge. The effective Hamiltonian for the system may be written as.

$$H = E_c\left(n - \frac{q_t^2}{e}\right) - 2E_J cos(\vartheta)cos\left(\frac{\Phi}{2\Phi_0}\right),\tag{3.28}$$

where ϑ is the phase difference between the two Josephson junctions in the SQUID and Φ is the quantized internal flux contained within the SQUID. In the circuit of Fig. 3.9, there is a current path from the bias voltage V_b through junctions 2 and b by which the phase across junction 2 (and therefore the phase difference ϑ) may be adjusted. The quantity q_t (the gate induced charge) is controlled by voltages V_g and V_P. Do we need both of these voltages? The answer given by Sect. 3.4 is "no" until we consider the qubit operation. The voltage V_g may be a steady voltage used to adjust the charge in the box, while V_P may be an oscillatory signal to induce Rabi oscillations between the two qubit states or a pulse to change the qubit state (or both) as discussed in Sect. 2.2. In [20], a pulse is used. The ability to adjust the qubit by the various voltages give a fine control over its operation.

One method of coupling charge qubits together lies in an external magnetic field, likely arising from a microwave signal and (perhaps) resonator in which the magnetic field external (to the SQUID) couples multiple qubits and induces correlations in the quantities V_P and Φ for each qubit with its neighbors [29].

Basic models of the charge qubit using multiple gates have been developed over the years [30]. These models have been extended to using a pair of charge qubits to form a single effective qubit [31, 32]. While the single SQUID charge qubit has been shown to work well, recent efforts have looked at coupling between two charge qubits, examining how they interact and can work cohesively together [29], and this has been extended to more complex combinations [33, 34].

The Transmon Qubit. When the charge qubit is shunted by a capacitor, as a method of reducing current noise, the device has been called a *transmon* [35]. If it were as simple

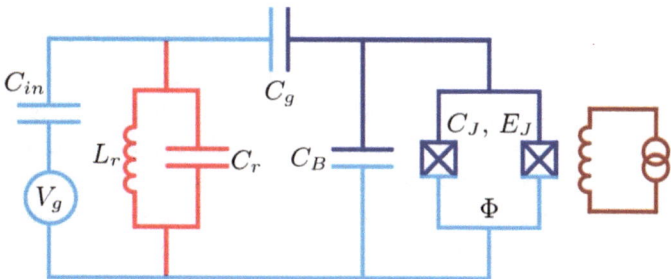

Fig. 3.10 The circuitry of a transmon qubit. The dark blue section forms the Cooper pair box, while the transmission line is denoted by the red inductor and capacitor. The circuit on the right couples inductively to the SQUID circuit. Reprinted with permission from [35], copyright 2007 by the American Physical Society

as that, there would not be much recognition. However, the shunt capacitor also reduces the charging energy and hence increases the size of the Josephson energy relative to the charging energy, so that we have $E_J > E_c$. This puts this ratio between that of the charge qubit and the flux/phase qubit (discussed below). Although this takes us away from a simple charge qubit, the transmon is often called a charge qubit because the origins lie in the presence of the Cooper pair box of Fig. 3.9. This may be seen in Fig. 3.10.

Technically, the transmon is short for a "transmission-line shunted charge qubit" in which plasma oscillations are important [35]. In Fig. 3.10, the SQUID is easily picked out on the right-hand side of the figure, while that part composing the Cooper pair box is the set of dark blue lines between the various capacitors and Josephson junctions. The transmission line is represented by the inductor-capacitor circuit in red. The qubit and the SQUID are also in parallel with the inductively-coupled circuit on the far right, which may operate in the microwave range, but certainly applies an oscillating magnetic field to the SQUID. It may also be noted that only a single oscillatory voltage source is present here (located on the far left).

The advantage of this qubit is based upon the fact that the dispersion in the box charge is quadratic as may be inferred from (3.23), but the various energy levels have an anharmonicity that is only linear in this charge. The anharmonicity is quite important, and may be defined from the various energy levels. First the Hamiltonian for the qubit remains only a slight variation of (3.27) (the flux in the SQUID is included in the definition of the Josephson energy E_J here). This Hamiltonian can be solved in a basis set of Mathieu[2] functions that yield energy levels as [36–38]

$$E_m(n_g) = E_c a_2(-2E_J/E_c). \tag{3.29}$$

---

[2] The Mathieu equation is a particular boundary eigenvalue equation that arises when the potential term has a cosinusoidal form.

where $a_n(Q)$ is a coefficient that arises in the solution of Mathieu's equation. Normally, these solutions yield energy *bands* that are typical in condensed matter physics, where the periodicity of the crystal can be represented by a cosinusoidal potential. For a sufficiently large value of the ratio E_J/E_c, these bands flatten out into discrete energy levels, giving atomic-like levels. It may be recalled from Chap. 2 that the harmonic oscillator (which has a quadratic or harmonic potential) gives equally spaced levels, and these are detrimental to keeping long coherence times, since the desired two levels easily couple to all other levels. Hence, it is desired to have anharmonic spacing of the levels, which may be defined by an anharmonicity factor.

$$\alpha = \frac{E_i - E_{i-1}}{E_0} \neq integer, i \geq 1. \tag{3.30}$$

Any integer value that results from this expression means the potential is harmonic (it is 2 in the harmonic oscillator, because the spacing is $\hbar\omega_0$ while $E_0 = \hbar\omega_0/2$). Another way of defining the anharmonicity is through the energy differences, as.

$$\alpha = \frac{E_{j,j-1} - E_{j+1,j}}{E_{10}}. \tag{3.31}$$

In this case, α will be zero for the harmonic oscillator, since all levels are equally spaced. In atom-like situations, $\alpha > 0$ as the levels get closer together as one goes up in energy. For hard-wall quantum wells, on the other hand, $\alpha < 0$ as the energy levels get further apart as one goes up in energy. (The harmonic case is the transition between these latter two behaviors.)

Optimizing all the various parameters from the circuit and the solutions to the Mathieu equation suggest that the Josephson energy and the charging energy should be in the range $20 < E_J/E_c < 5 \times 10^4$ [35]; that is, somewhere between the charge qubit and the flux/phase qubit which often has the larger value. With the charge n_g normalized to the Cooper pair charge of $2e$, and for a sufficiently large ratio of the two energies, the requisite qubit energy levels may be written as [35]

$$E_m = E_m(n_g = 1/4) - \frac{\varepsilon_m}{2}cos(2\pi n_g)$$
$$\varepsilon_m = E_m(n_g = 1/2) - E_m(n_g = 0), \tag{3.32}$$

and these have the asymptotic limits for a large energy ratio [35]

$$\varepsilon_m = (-1)^m E_c \frac{2^{4m+5}}{m}\sqrt{\frac{2}{\pi}}\left(\frac{E_J}{2E_c}\right)^{\frac{m}{2}+\frac{3}{4}} exp\left(-\sqrt{\frac{8E_J}{E_c}}\right).$$
$$E_m = -E_J + \hbar\omega_p\left(m + \tfrac{1}{2}\right) - \frac{E_c}{12}\left(6m^2 + 6m + 3\right) \tag{3.33}$$

Equation (3.32) just asserts the previous comment that the Mathieu equation yields energy bands that are cosinusoidal in nature, replicating the real-space potential. For this reason, these solutions often appear in condensed matter physics. One notes from (3.33) that the amplitude of the oscillator term is exponentially damped by the ratio of the two energies. The quantity $\omega_p = \sqrt{8E_JE_c}/\hbar$ is the Josephson plasma frequency. As this is quantized, excitation of this involves the emission or absorption of plasmons with the energy $\hbar\omega_p$. The transmission line indicated by the red part of the circuit actually creates a resonator in its own right, as shown by the inductor and capacitor in parallel. Thus, if the SQUID-box circuit is to act as a two level atom, the resonator provides the same function as in the Jayne-Cummings model of Chap. 2. So, as mentioned earlier, this model has significant application in the transmon.

Suppressing the noise fluctuations in the transmon is the basic rationale for its use as a qubit. The actual fact that the noise is suppressed has been clearly shown in experimental systems [39]. More recently, it has been shown that relaxation time fluctuations, which arise partially from the noise fluctuations, have a power law scaling [40]. One problem that can arise is that memory effects have been detected in some conditions of operation [41].

The fact that the qubit operates with the emission or absorption of a photon at the plasmon frequency, means that it is possible to operate with single photon processes in the microwave range [42]. Variations in the coupling between the transmission line resonator and the charge qubit can be dependent upon how the two are coupled, and this is a concern in circuit design [43]. Nevertheless, these qubits can be operated at a very low signal level, even at the threshold for which quantum error corrections can be achieved [44].

The coupling of the system on the right-hand side of Fig. 3.10 just points out that measurement and control of the box qubit is commonly done by means of microwave resonators with the techniques usual in quantum electrodynamics, as in other superconductor qubits [45]. Actual control of a transmon qubit means suppressing spontaneous emission processes, which can arise or be affected by off-resonant modes of the cavity/resonator [46]. Nevertheless, digital control can be achieved using pulsed signals [47, 48]. Speedup of operation can be achieved by the use of bipolar single-flux quanta pulse sequences [49]. However, it is well-known that any harmonic oscillator can be driven into chaos by large enough pumping signals. Similarly, one has to be concerned that driving these circuits too hard can also lead to chaotic behavior [50]. Nevertheles, it is possible to couple a pair of transmons nonlinearly to photons and produce nonlinear optical effects such as photon blockade [51]. Despite a variety of concerns, one can remark that the noise suppression has been effective and transmons have become mainstream in the qubit world [52].

It is possible to create the transmon with a single Josephson junction (much as a single junction SQUID is possible, and an example is shown in Fig. 3.11 [53, 54]. In making real circuits, connecting the qubit with microwaves in a physical cavity is often in three-dimensions, and this distinction has appeared in the literature [55]. More recently, the

Fig. 3.11 Schematic diagram of the single Josephson junction and quantum dot that form a transmon. Reprinted from [54] under the creative commons 4.0 license

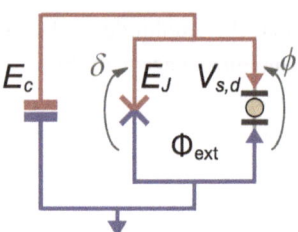

transmon has been integrated with silicon technology through the use of a superconductor-semiconductor hybrid approach [56, 57]. One advantage of the transmon is that it has a long coherence time, that can approach 0.5 ms [58].

The flux/phase qubit. An interesting aspect of electrical circuits is that a resonant circuit, such as the one in Fig. 3.10, can be a quantum harmonic oscillator [59]. For this simple circuit, the Hamiltonian can be written as

$$H = \frac{Q^2}{2C} + \frac{\Phi^2}{2L}. \tag{3.34}$$

Here, the charge Q and the flux Φ are conjugate variables in the same sense as position and momentum in classical mechanics. In quantum mechanics, these two variables become non-commuting operators. This will become important later, but this is one major method of coupling electromagnetic flux to the SQUID. In the flux/phase qubit, both the qubit and the box charge are retained, but the ratio of energies satisfies $E_J/E_c \gg 1$, and may be larger than 5×10^4, as discussed above with the transmon qubit. This just means that the charging energy has to be very small. The eigenstates of the SQUID ring represent two counter-rotating persistent currents, corresponding to a fixed number of flux quanta (h/e) contained in the ring. The inductance of this ring gives rise to a parabolic potential, as seen in (3.15), and adding the Josephson junction oscillating potential causes a nonlinearity that provides anharmonicity to separate the lowest states from higher lying levels. A typical flux/phase qubit is shown in Fig. 3.12 [60]. Using the two lowest levels of the qubit gives rise to a Hamiltonian that can be written as [18]

$$H = 4E_C n^2 + E_L(\varphi - \varphi_e)^2 - E_J cos(\varphi), \tag{3.35}$$

where $E_L = \Phi_0^2/2L$ is the inductive energy, $\varphi = 2\pi \Phi/\Phi_0$ remains the Josephson phase within the SQUID, and φ_e is the external flux. In Fig. 3.12, panel (a) shows a single Josephson junction in the SQUID ring, and the schematic circuit is depicted in panel (b). Panel (c) depicts the effective potential in the system as a function of the flux in the ring, for an external flux of $\Phi_0/2$ (a condition that is discussed more in Sect. 3.3 above). With this external flux, the potential is a symmetric double well at the very bottom with $|L\rangle$ and $|R\rangle$ as the two counter-rotating states mentioned above. These two states will interact to create "bonding" and "anti-bonding" hybrids of these two states (the anti-bonding

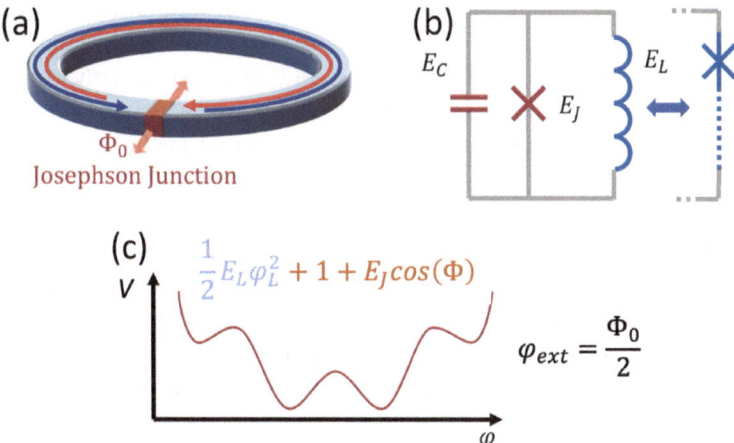

Fig. 3.12 An rf SQUID qubit. **a** The simplest flux qubit composed of a single Josephson junction SQUID. **b** The equivalent circuit of the qubit in which the SQUID (junction and inductor) is shunted with a capacitor. **c** Total potential when the external magnetic field is one-half of Φ_0. Reprinted with permission from [60], with permission of AIP Publishing

state lies at a higher energy than the bonding state, thus creating the two levels of the qubit). These latter two states arise from the macroscopic tunneling through the Josephson junction, a process that causes a coupling interaction between the two counter-rotating current states.

It has been shown that this type of flux qubit can have a long coherence time, with a relaxation time on the order of a few microseconds [61]. This depends upon controlling the critical inductance in the ring [62], and controlling the gap that opens between the two levels used in the qubit [63].

This flux qubit system was investigated as to the role of dissipation in the system [64]. In this latter case, the dissipation is incorporated by coupling the system to a reservoir, and this expanded system is then projected back onto the reduced density matrix for the ring-junction system. They conclude that the two coherent states survive even in the presence of dissipation, at least for weak dissipation. In some sense this is limited by the noise in the system, which can consist of both flux and charge noise [65], although there is a component of $1/f$ noise as well [66].

Other Forms of Superconducting Qubits. While the above discussions describe the three main types of superconducting qubits, engineers and scientists are always tinkering with these designs. This has led to a number of variants that have appeared in the literature. Not all of these can be discussed, but a few warrant description, as they tend to be operated in interesting regimes. One of these is the *unimon*, whose circuit appears exactly as in Fig. 3.12b. However, the unimon is operated under the conditions of not only $E_J/E_c \gg 1$, but also $E_L/E_c \gg 1$ [67]. Similarly, the fluxonium qubit is described as a inductively coupled qubit, quite similar to the unimon, but with lower values $E_L/E_c \sim 1$. While some versions use the same circuit as Fig. 3.12b [68], others add additional capacitors and inductors within the SQUID loop [69]. Still others move further away from the unimon by just describing the circuit as an inductively coupled transmon [70]. The main purpose of all of these variants is to increase the coherence time by reducing the noise in the qubit.

A second main variant is referred to as a charge-flux hybrid qubit. As noted above, most transmon qubits use a single Josephson junction in the SQUID for r.f. purposes. If a second Josephson junction is added within the ring, one can isolate the charge box between the two junctions, and still couple flux through the ring. If the Coulomb energy E_C dominates the Josephson energy, the charge qubit limit results. If the Josephson energy dominates the Coulomb energy, the flux qubit limit is reached. But, if the two energies are comparable with $E_J \geq E_C$, one can consider the qubit to be in a charge-phase, or hybrid charge-flux qubit regime [18]. Just as described above, the system can still act as a two-level atom, and sometimes is referred to as a quantronium [21]. One version of the quantronium thus adds two Josephson junctions in series in a path that shunts the circuit of Fig. 3.12b. These two junctions create the charge box, but the ring from these to the larger third junction (the junction already in the figure) creates the supercurrent, and the flux reservoir [71]. As usual, external flux will vary the flux enclosed in this ring.

3.6   Qubit Circuits

The development of quantum information systems now requires one to move from the qubit level to that of a gate level, which can contain several qubits. In order to move beyond the level of a single isolated qubit, many things must be achieved. First, it is absolutely necessary that a long lifetime must be achieved for the qubit, since the state of the qubit must survive long enough for it to manipulated or read during gate operations [72, 73]. In addition, the DiVincenzo requirements [74], discussed in Sect. 2.4, must be met for the circuit. Above the individual qubit level, one must also be able to implement quantum error correction to overcome the problems of the qubits (especially with respect to noise) that have been discussed above. And, in many cases, some method of feedback to the qubit may be necessary to stabilize it at a level necessary to hold the information for longer times. In Fig. 3.13, a conceptual view of the layers of information and control necessary for a quantum information processor is shown [75]. It may be seen that as one flows downward from the algorithm, or program level, to the qubit level, the requirements transition from programming to physical science, in which the previous sections above deal with this lower level. But, the problems inherent in the qubit necessitate the intermediate levels of qubit control and quantum error correction.

The development of single qubit gates [76, 77] and two-qubit gates has allowed this process to become real over the recent decade(s). Often, the qubits used in various chips

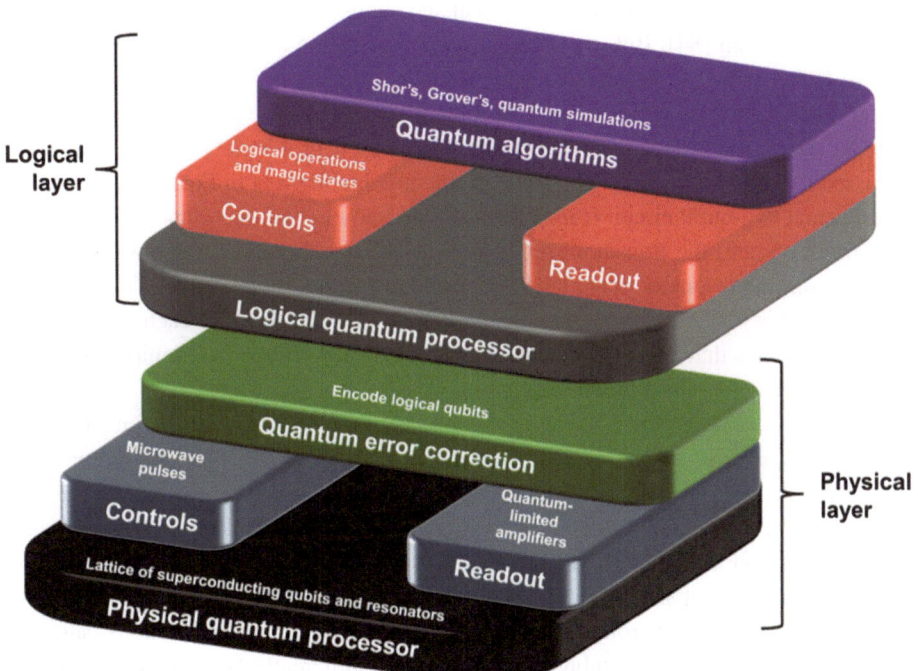

Fig. 3.13 An overview of a total processor system. The physical and logical levels are separated by other levels necessary for qubit control and quantum error correction. These lower levels are, in turn connected to the global control and algortihm level. Reprinted from [75] under the creative commons 4.0 license

are based upon the transmon qubit described above. In the following paragraphs, various gate circuits based upon one, two, or more qubits will be described in further detail. But, the reader should remember that operating quantum processors exist today (discussed in Chap. 1). Thus, the descriptions given below should be recognized as examples of how real gate operations may be established in circuits using real qubits.

Single Qubit Gates. Single qubits have been made into gates by capacitive coupling, for example, a charge qubit coupled to a high-quality superconducting resonator [78], similar in concept to the Jaynes-Cummings model. Here, the strength of the coupling is adjusted by the size of the capacitive coupler. A similar approach has used inductive coupling between the qubit and the resonator [79]. And, a single transmon qubit has been used in a "three-dimensional" gate by placing the qubit into a cavity resonator providing direct coupling via a fast-flux line (which is what seems to be basically an antenna type of coupling to the cavity resonator) [80].

Approaches to improving coherence time in single qubit devices have involved for example closed-loop feedback [81] with an external optimal control strategy [82]. The nature of the 1/f noise in a strongly coupled qubit-resonantor system has been studied

[83], and an error correction method for leakage control has also been developed for the single qubit system [84]. But, single qubit gates are not the most needed type of circuit.

Two Qubit Gates. Coupling multiple qubits together to create gates is an important part of building quantum information systems. This has already been pointed out above. But, the actual coupling of two qubits can be achieved in a variety of ways, as will be discussed a bit in this section. The basic problem is highlighted in Fig. 3.14, which is an experimental two-qubit circuit from IBM used for studying the important role of coupling [75]. In this figure, the transmon qubits are located at either side of the figure, while a transmission line resonator occupies the central part of the figure. Each of the transmons has a readout resonator as well. The lines running back and forth in each region are the transmission lines that can be either inductor or resonator. This is an example of a circuit with two qubits, a connecting bus and two read-out resonators.

The above example can exist on a single chip, and this chip can be integrated into the system in a variety of ways. The actual package may contain a single chip or several chips. That is, several chips can be packaged together to create a processor in a single package. The package itself is connected to various power supplies, microwave sources and detectors, and other control signal lines. In some approaches chips may be flipped to bond to the package upside down (flip-chip bonding) [85].

Sometimes in moving from qubits to gates, unwanted circuit interactions can occur. For example, often transmon qubits do not possess a sufficiently large anharmonicity, and

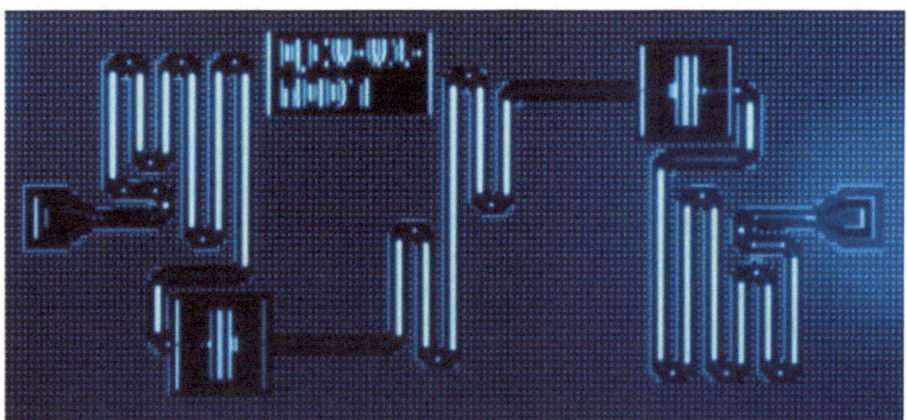

Fig. 3.14 An example of a circuit containing two qubits (lower left and upper right), a coupling bus (center), and two read-out resonators (one for each qubit, at upper left and lower right). Reprinted from [75] under the creative commons 4.0 license

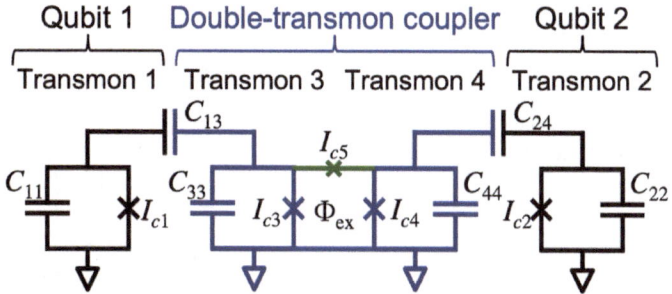

Fig. 3.15 A double-transmon coupler (in blue) is capacitively coupled to the two logic qubits (in black). This two transmon coupler effectively eliminates the residual ZZ interaction. Reprinted from [96] under the Creative Commons 4.0 license

this is suspected of leading to unwanted ZZ gate[3] action [86]. This interaction has actually been stimulated, and then cancelled to create a CZ gate[4] [87].

One method of coupling qubits is simple direct coupling. This can be done with a pair of SQUID based qubits, or even three such qubits, merely by them having a common edge of the loops that is shared between the two qubits [88]. Another method has created a semiconductor-superconductor quantum dot spin qubit which is then embedded within a transmon qubit to provide this direct coupling [89].

The generation of coherence and entanglement between two charge qubits that are capacitively coupled has been studied [32]. Similarly, a circuit containing two transmon qubits that are capacitively coupled, and then connected to two control microwave signals, also has been studied [90]. And, capacitive coupling between a pair of transmons, when one is tunable and the other parity protected, has been achieved [91].

Coupling of two qubits can also be done by using an additional qubit as the coupler, the latter of which is capacitively coupled to the first two qubits. This can be done for transmons by using either one [92–94] or two transmons in the coupling circuit [95, 96]. An example of the latter is shown in Fig. 3.15. Here, two fixed-frequency transmons that are coupled through a two transmon circuit (in blue in the figure) containing an additional Josephson junction. This part of the circuit provides a tunable coupler between the two logic qubits at either side (black in the figure). The coupling between the two logic qubits is controlled by the magnetic field that passes through the central loop in the coupler (where the external flux Φ_e is indicated). One advantage of this type of coupler is that the coupling needed to produce a ZZ gate, noted above, vanishes for detuned qubits. This is an advantage over the single transmon coupler. Similar approaches have also been used for fluxonium qubits [97].

---

[3] The ZZ gate is often written as $R_{zz}(\phi)$ and is a two-qubit rotation, with both being rotated around the z-axis.

[4] The CZ gate is a one qubit controlled rotation around the z-axis.

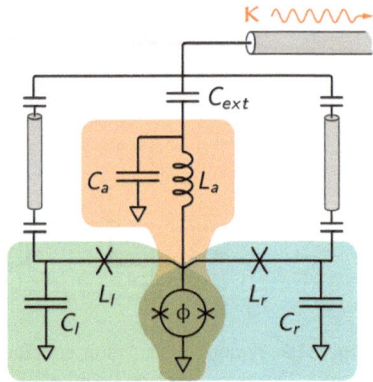

Fig. 3.16 A resonator coupled
gate in which two transmons
are coupled to a circuit
resonator (in orange).
Reprinted from [98] under the
Creative Commons 4.0 license

Coupling between qubits can also be achieved by using resonators or resonant circuits [98]. Such a circuit is shown in Fig. 3.16 for a resonator coupling between a qutrit and a qubit with transmon circuits. The two transmon circuits are shown in green and blue and the coupler is shown in orange. The coupler and the transmons share a common SQUID that implements the coupling through a parametric interaction. The read out resonators are indicated in grey.

One of the oldest methods of coupling two qubits is through the use of a cavity resonator composing a signal bus (already shown in Fig. 3.14) [78, 99–101]. A schematic of the qubit resonator coupling is shown in Fig. 3.17. The resonator is formed by a finite length transmission line, and the two qubits are embedded within the superconducting transmission line. Shown below the schematic is the lumped element equivalent circuit with the qubits indicated in green.

Finally, two qubits may be coupled by an additional Josephson junction and capacitor [102]. Here, the two qubits are flux qubits using three Josephson junctions in each.

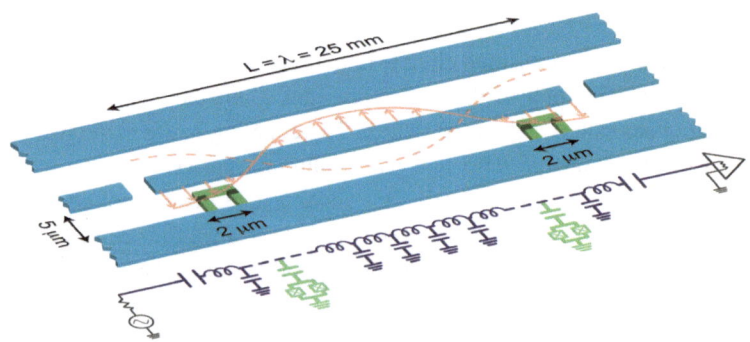

Fig. 3.17 Layout and the lumped parameter version (below) of resonator and two qubits (green). These qubits are formed inside the superconducting transmission line resonator (blue). Reprinted with permission from [78], copyright 2007 by the American Physical Society

Multiple Qubit Gates. The time evolution of an entangling gate that requires only a single step to produce the entanglement of three transmon qubits at GHz frequencies has been measured [103]. Here, the three transmon gates are capacitively coupled to a resonator that is driven by a strong microwave field. The resulting entanglement is characterized using tomographic methods. Earlier, it had been shown that the resonator coupled to three qubits could be used to both read out and reset the qubits [104]. In this latter work, a parametric interaction between the lossy resonator and the three tunable superconducting qubits gives the transfer of the information from the qubits to the resonator and also serves as the readout. This transfer resets the qubits as well.

For the past few years, the field has grown to where real circuits are appearing in real machines and experiments to verify the performance of those circuits and machines. More complicated qubit and circuit dynamics is often required for accurate performance [105]. In Fig. 3.18, a five qubit circuit with the connecting transmission line resonators is shown [106]. Here, four of the qubits are used for encoding two logical qubits and the fifth qubit (the central qubit) is a monitoring qubit (sometimes called a syndrome qubit in analogy to monitoring a disease). The qubits are coupled by coplanar waveguides (the transmission lines), and each qubit has its own resonator which is also used for readout. Circuits such as these can be effectively mounted on circuit boards in a manner that still allows access for microwave signals. An example is shown in Fig. 3.19 [107]. The chip is mounted through a hole cut in the printed circuit board of the computer, and a so-called sub-board, or secondary circuit is mounted below the circuit board. This sub-board carries the signal lines so must be microwave compatible. The actual circuit is then flip-chip mounted on this sub-board. This approach is used in IBM's newer "Eagle" processors [107].

Finally, in Fig. 3.20, is shown a more complicated line of qubits on a single chip. Here, an array of 24 transmon qubits, based on sub-100 nm Josephson Junctions, each of which has a microwave coplanar resonators, and leads to control signals. Interactions among the qubits occurs through the resonators. As remarked in Chap. 1, today there are machines available with hundreds of qubits.

Fig. 3.18 Color enhanced micrograph of a five qubit lattice, with interconnecting transmission lines. Reprinted with permission from [106], copyright 2017 by the American Physical Society

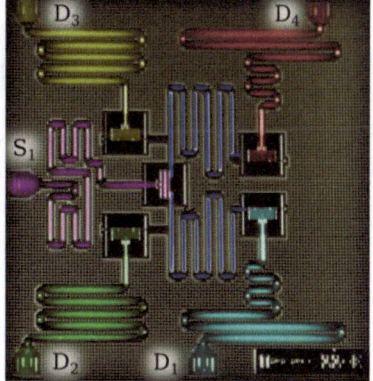

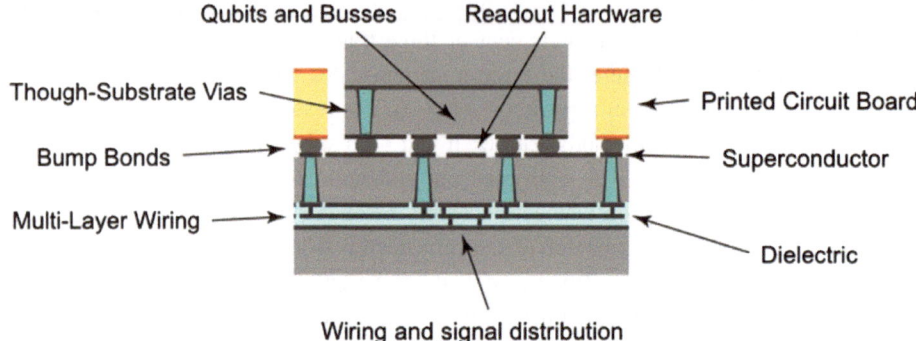

Fig. 3.19 An example of a scheme that allows breaking the plane of the circuit board for signal delivery. Here, the actual chip, such as that in Fig. 3.18, is flip-chip mounted on a sub-board (that is, the chip is turned upside down, so that the active circuits are now on the bottom as the chip is mounted to the board), located below the actual printed circuit board (in yellow). This sub-board has the signal lines that are fed to the actual chip. Reprinted from [107], with permission of AIP Publishing

Fig. 3.20 An array of 24 transmon qubits, with their individual resonators and the external lines for bias and control. Reprinted from Wikimedia Commons under the Creative Commons Attribution 4.0 International

3.7   Superconducting Sensors

Sensors composed of superconducting materials are among the oldest quantum sensors. These can be either formed aroung the SQUID or the SET. The largest use of the SET for sensing is with semiconducting materials, and this discussion is left to Chap. 5. Sensing with SQUIDs is based upon their extreme sensitivity to magnetic fields. As was discussed above, as the external magnetic field is increased, the internal flux increases in jumps of $\Phi_0 = h/e$, which also leads to increments in the voltage across the circuit. Hence, counting the flux jumps leads to an accurated digitized measurement of the external magnetic field. This has led to applications in defense, medical, and biological fields as well as electronics.

The use of magnetic sensing seems to date to the study of the earth's magnetic field and geoexploration as early as the mid-nineteenth century. Since large iron and steel objects tend to develope their own magnetic fields, the U. S. Navy has used magnetometers to search for submarines since World War II. Before that, magnetometers were used in oil and gas exploration. By the 1970s, the use of SQUIDs and Josephson junctions was well entrenched technology for the search for "magnetic anomalies," variations in the earth's magnetic field caused by large steel submerged objects; e.g., submarines. The normal search involves low flying aircraft cycling back and forth across the ocean.

The use of magnetometers to detect the magnetic fields associated with heart and brain currents dates to the 1960s [108, 109], and by 1970, the Josephson junction had been adopted for the detector [110]. Today, the SQUID is used in these fields as well as for low magnetic field magnetic resoance imaging (MRI) [111]. One could say that quantum sensing is well established in the medical and biological world. This is especially true if the SQUID is mounted on an atomic force microscope, in which the deflection of a beam cantilever due to external force on the beam is measured [112]. Use of the SQUID gives an ultrasensitive detector that has extremely high resolution (of course, depending upon the distance from the detector to the source). In some usage, the SQUID participates as a transition-edge sensor, where the temperature is near the transition temperature of a phase transition such as that associated with the onset of superconductivity. Often it is not the SQUID that goes through the transition, but rather it is used as a current amplifier. That is, the SQUID operates as a readout device [113].

Currently, these SQUID devices can be used for accelerometers [114] and ammeters [115]. And, of course, when the SQUID is used as, or with, a qubit, this also provides a viable quantum sensor [116]. But, the days of sensing with SQUIDs may be drawing to a close as more modern approaches such as giant magnetoresistance devices are intruding into this world.

References

1. van Delft, D.: Freezing Phyics: Heike Kamerlingh Onnes and the Quest for Cold. Tr. by Jackson, B. (Amsterdam, Royal Netherlands Academy of Arts and Science, 2007).
2. Bardeen, J., Cooper, L., Schriffer, J. S.: The Theory of Superconductivity. Phys. Rev. **108**, 1175 (1957).
3. Cooper, L.: Bound Electrons in a Degenerate Fermi Gas. Phys. Rev. **104**, 1189 (1956).
4. Josephson, B. D.: Possible New Effects in Superconductive Tunneling. Phys. Lett. **1**, 251 (1962).
5. Josephson, B. D.: The Discovery of Tunneling Supercurrents. Rev. Mod. Phys. **46**, 251 (1974).
6. Shapiro, S.: Josephson Currents in Superconducting Tunneling: The Effects of Microwaves and Other Observations. Phys. Rev. Lett. **11**, 80 (1963).
7. Riedel, E.: Zum Tunneleffeckt bei Supraleitern im Mikrowellenfeld. Z. Naturforsch. **19a**, 1634 (1964).
8. Hamilton, C. A., Shapiro, S.: Experimental Demonstration of the Riedel Peak. Phys. Rev. Lett. **26**, 426 (1971).
9. Gorter, C.: A Possible Explanation of the Increase of the Electrical Resistance of Thin Metal Films at Low Temperatures and Small Field Strengths. Physica **17**, 777 (1951).
10. Dramois, E.: Interpretation of the Work Function Value in very Thin Metallic Films. J. Phys. Radium **17**, 210 (1956).
11. Averin, D. V., Likharev, K. K.: Coulomb Blockade of Single-Electron Tunneling, and Coherent Oscillations in Small Tunnel Junctions. J. Low Temp. Phys. **62**, 345 (1986).
12. Oh, S., Choi, S.-I.: Wigner-function approach to a Single-Electron Tunnel Junction. Phys. Rev. B **54**, 4440 (1996).
13. Jaklevic, R. J., Lambe, J., Silver, A. H., Mercereau, J. E.: Quantum Interference Effects in Josephson Tunneling. Phys. Rev. Lett. **12**, 159 (1964).
14. Fultan, T. A., Dolan, G. J.: Observation of Single-Electron Charging Effects in Small Tunnel Junctions. Phys. Rev. Lett. **59** 109 (1987).
15. van Bentum, P. J. M., van Kempen, H., van de Leemput, L. E. C., Teunissen, P. A. A.: Single-Electron Tunneling Observed with Point Contact Tunnel Junctions. Phys. Rev. Lett. **60**, 369 (1988).
16. Ferry, D. K.: Transport in Semiconductor Mesoscopic Devices, 2nd ed. (Bristol, IOP Publishing, 2020) chapter 8.
17. Liu, Y.-X., Wei, L. F., Nori, F.: Measuring the Quality Factor of a Microwave Cavity using Superconducting Qubit Devices. Phys. Rev. A **72**, 033818 (2005).
18. Wendin G.: Scalable Solid-State Qubits: Challenging Decoherence and Read-out. Phil. Trans. Roy. Soc. London A **361**, 1323 (2003).
19. Mooij, J. E., Orlando, T. P., Levitov, L., Tian, L., et al.: Josephson Persistent-Current Qubit. Science **285**, 1036 (1999).
20. Nakamura, Y., Pashkin, Yu. A., Tsai, J. S.: Coherent Control of Macroscopic Quantum States in a Single-Cooper-Pair Box. Nature **398**, 786 (1999).
21. Vion, D., Aassime, A., Cottet, A., Joyez, et al.: Manipulating the Quantum State of an Electrical Circuit," Science **296**, 886 (2002).
22. Steffen, M., DiVincenzo, D. P., Chow, J. M., Theis, T. N., Ketchen, M. B.: Quantum Computing: An IBM Perspective," IBM J. Res. Develop. **55**, 13 (2011).
23. Park, S. H., Baek, D., Park, I., Hahn, S.: Scaling of Superconducting Quantum Circuit using Flip-Chip Assembly. IEEE Trans. Appl. Supercond. **33**, 1701206 (2023).

24. Pishchimova, A. A., Smirnov, N. S., Ezenkova, D. A., Krivkov, E. A., *et al.*: Improving Josephson Junction Reproducibility for Superconducting Quantum Circuits: Junction Area Fluctuation. Sci. Rpts. **13**, 6772 (2023).
25. Lisenfeld, J., Bilmes, A., Ustinov, A. V.: Enhancing the Coherence of Superconducting Quantum Bits with Electric Fields. NPJ Quantum Inform. **9**, 8 (2023).
26. Giustina, M.: Superconducting Qubits Cover New Distances. Nature **617**, 254 (2023).
27. Arute, F., Arya, K., Rabbash, R., Bacon, D., *et al.*: Quantum Supremacy Using a Programmable Superconducting Processor. Nature **574**, 505 (2019).
28. Bouchiat, V., Vion, D., Joyez, P., Esteve, D., Devoret, M. H.: Quantum Coherence with a Single Cooper Pair, Phys. Scripta T **76**, 165 (1998).
29. Zhang, C., Chan, G. X., Wang, X., Xue, Z.-Y.: Coupling Two Charge Qubits via a Superconducting Resonator Operating in the Resonate and Dispersive Regime. Phy. Rev. A **106**, 032608 (2022).
30. Berrada, K., Abdel-Khalek, S., Algarni, M.: Coherence, Purity, and Correlation for Superconducting Charge Qubits. Results Phys. 48, 106414 (2023).
31. Naveena, P., Muthuganesan, R., Chadrasekar, V. K.: Effects of Decoherence on Quantum Correlations in a Two Superconducting Charge Qubit System. Physica A **592**, 126852 (2022).
32. Benzahra, M., Mansour, M., Oumennana, M., Elghaayda, S.: Quantum Correlations and Thermal Coherence in a Two-Superconducting Charge Qubit System. Laser Phys. **33**, 075202 (2023).
33. Yu, J., Retamal, J. C., Sanz, M., Solano, E., Albarran-Arriagada, F.: Superconducting Circuit Architecture for Digital-Analog Quantum Computing. EPJ Quantum Technol. **9**, 9 (2022).
34. Shamir, S., Khan, M. A., Abbas, T., Alvi, S. H., Islam, R.: Mutli-Particle Entanglement Generation Through Superconducting Circuits. Int. J. Theor. Phys. **62**, 102 (2023).
35. Koch, J., Yu, T. M., Gambetta, J., Houck, A. A., *et al.*: Charge-Insensitive Qubit Design Derived fom the Cooper Pair Box. Phys. Rev. A **76**, 042319 (2007).
36. Brillouin, L.: Wave Propagation in Periodic Structures. (New York, McGraw-Hill, 1946).
37. Morse, P. M., Feshbach, H.: Methods of Theoretical Physics. (New York, McGraw-Hill, 1953).
38. Abramowitz, M., Stegun, I. A.: Handbook of Mathematical Functions. (Washington, National Bureau of Standards, 1964).
39. Schdreier, J. A., Houck, A. A., Koch, J., Schuster, D. K., *et al.*: Suppressing Charge Noise Decoherence in Superconducting Charge Qubits. Phys. Rev. B 77, 180502 (2008).
40. Li, K., Datta, S. K., Steffen, Z., Palmer, B. S., *et al.*: Power-Law Scaling of Relaxation Time Fluctuations in Transmon Qubits. IEEE Trans. Appl. Supercond. **33**, 1700205 (2023).
41. Gulácsi, B., Burkard, G.: Signatures of Non-Markovianity of a Superconducting Qubit. Phys. Rev. B **107**, 174511 (2023).
42. Houck, A. A., Schuster, D. I., Gambetta, J. M., Schreier, J. A., *et al.*: Generating Single Microwave Photons in a Circuit. Nature **449**, 328 (2007).
43. Devoret, M., Girvin, S., Schoelkopf, R.: Circuit QED: How Strong can the Coupling Between a Josephson Junction Atom and a Transmission Line Resonator Be? Ann. Phys. **16**, 767 (2007).
44. Chen, L., Li, H.-X., Lu, Y., Warren, C. W., *et al.*: Transmon Qubit Readout Fidelity at the Threshold for Error Correction without a Quantum-Limited Amplifier. NPJ Quantum Inform. **9**, 26 (2023).
45. Zhu, M.-Z, Ye, L.: Implementing Phase-Covariant Cloning in Circuit Quantum Electrodynamics. Ann. Phys. **373**, 512 (2016).
46. Houck, A. A., Schreier, J. A., Johnson, B. R., Chow, J. M., *et al.*: Controlling the Spontaneous Emission of a Superconducting Transmon Qubit. Phys. Rev. Lett. **101**, 080502 (2008).

47. Howe, L., Castellanos-Beltran, M. A., Sirois, A. J., Olaya, D., *et al.*: Digital Control of a Superconducting Qubit Using a Josephson Pulse Generator at 3 K. PRX Quantum **3,** 010350 (2022).

48. Castellanos-Beltran, M. A., Sirois, A. J., Howe, L., Olaya, D., *et al.*: Coherence-Limited Digital Control of a Josephson Pulse Generator at 3 K. Appl. Phys. Lett. **122,** 192602 (2023).

49. Vozhakov, V., Bastrakova, M., Klenov, N., Satanin, A., Soloviev, I.: Speeding Up Qubit Control with Bipolar Single Flux Quantum Sequences. Quantum Sci. Techn. **8,** 035024 (2023).

50. Cohen, J., Petrescu, A., Shillito, R., Blais, A.: Reminiscence of Classical Chaos in Driven Transmons. PRX Quantum **4,** 020312 (2023).

51. Wang, Y., Zhang, G.-Q., You, W.-L.: Photon Blockaded with Cross-Kerr NonLinearity in Superconducting Circuits. Laser Phys. Lett. **15,** 105201 (2018).

52. Houck, A. J., Koch, J., Devoret, M. H., Girvin, S. M., Schoelkopf, R. J.: Life After Charge Noise: Recent Results with Transmon Qubits. Quantum Inform. Proc. **8,** 105 (2009).

53. Place, A. P. M., Rodgers, L. V. H., Mundada, P., Smitham, B. N., et al.: New Materials Platform for Superconducting Transmon Qubits with Coherence Times Exceeding 0.3 Millisecond. Nature Commun. **12,** 1779 (2021).

54. Bargerbos, A., Pita-Vidal, M., Zitko, R., Avila, J., *et al.*: Singlet-Doublet Transitions of a Quantum Dot Josephson Junction Detected in a Transmon Circuit. PRX Quantum **3,** 030311 (2022).

55. Paik, H., Schuster, D. I., Bishop, L. S., Kirchmair, G., *et al.*: Observation of High Coherence in Josephson Junction Qubits Measured in a Three-Dimensional Circuit QED Architecture. Phys. Rev. Lett. **107,** 240501 (2011).

56. Hertel, A., Andersen, L. O., van Zanten, D. M. T., Eichinger, M., *et al.*: Electrical Properties of Selective-Area-Grown Superconducter-Semiconductor Hybrid Structures on Silicon. Phys. Rev. Appl. **16,** 044015 (2021).

57. Hertel, A., Eichinger, M., Andersen, L. O., van Zanten, D. M. T., *et al.*: Gate-Tunable Transmon Using Selective-Area-Grown Superconducter-Semiconductor Hybrid Structures on Silicon. Phys. Rev. Appl. **18,** 034042 (2022).

58. Wang, C., Li, X., Xu, H., Li, Z., *et al.*: Toward Practical Quantum Computers: Transmon Qubit with a Lifetime Approaching 0.5 Milliseconds. NPJ Quantum Inform. **8,** 3 (2022).

59. Ferry, D. K.: *Quantum Mechanics*, 3rd Ed. ((Inst. Phys. Publ., Bristol, 2021).

60. Dmitriev, A. Yu., Astafiev, O. V.: A Perspective on Superconducting Flux Qubits. Appl. Phys. Lett. **119,** 080501 (2021).

61. Chang, T., Cohen, T., Holzman, I., Catelani, G., Stern, M.: Tunable Superconducting Flux Qubits with Long Coherence Times. Phys. Rev. Appl. **19,** 024066 (2023).

62. Dausy, H., Nulens, L., Raes, B., Van Bael, M. J., *et al.*: Impact of Kinetic Inductance on the Critical-Current Oscillations of Nanobridge SQUIDS. Phys. Rev. Appl. **16,** 024013 (2021).

63. Chang, T., Holzman, I., Cohen, T., Johnson, B. C., *et al.*: Reproducibility and Gap Control of Superconducting Flux Qubits. Phys. Rev. Appl. **18,** 064062 (2022).

64. Zou, J., Shao, B., Su, W.-Y.: Wave-Packet Analysis of Mesoscopic Josephson Junction with Dissipation in the Wigner Formalism. Phys. Lett. A **285,** 401 (2001).

65. Whiticar, A. M., Smirnov, A. Y., Lanting, T., Whittaker, J., *et al.*: Probing Flux and Charge Noise with Macroscopic Quantum Tunneling. Phys. Rev. B 107, 075412 (2023).

66. Rower, D. A., Ateshian, L., Li, L. H., Hays, M., *et al.*: Evolution of 1/f Flux Noise in Superconducting Qubits with Weak Magnetic Fields. Phys. Rev. Lett. **130,** 220602 (2023).

67. Hyyppä, E., Kundu, S., Chan, C. F., Gunyhó, A., *et al.*: Unimon Qubit. Nature Commun. **13,** 6895 (2022).

68. Bao, F., Deng, H., Ding, D., Gao, R., *et al.*: Fluxonium: An Alternative Qubit Platform for High-Fidelity Operations. Phys. Rev. Lett. **129,** 010502 (2022).

69. Moskolenko, I. N., Besedin, I. S., Simakov, I. A., Ustinov, A. V.: Tunable Coupling Scheme for Implementing Two Qubit Gates on Fluxonium Qubits. Appl. Phys. Lett. **119**, 194001 (2021).

70. Hassani, F., Perruzo, M., Kapoor, L. N., Trioni, A., *et al.*: Inductively Coupled Transmons Exhibit Noise Insensitive Plasmon States and a Fluxon Decay Exceeding 3 Hours. Nature Commun. **14**, 3968 (2023).

71. Hutchinson, G. D., Holmes, C. A., Stace, T. M., Milburn, G. J., *et al.*: Model for an Irreversible Bias Current in the Superconducting Qubit Measurement Process. Phys. Rev. A **74**, 062302 (2006).

72. Schoelkopf, R. J., Girvin, S. M.: Wiring Up Quantum Systems. Nature 451, 664 (2008).

73. Devoret, M. H., Schoelkopf, R. J.: Superconducting Circuits for Quantum Information: An Outlook. Science **339**, 1169 (2013).

74. DiVincenzo, D. P.: The Physical Implementation of Quantum Computing. Fortschr. Phys. **48**, 771 (2000).

75. Gambetta, J. M., Chow, J. M., Steffen, M.: Building Logical Qubits in a Superconducting Quantum Computing System. NPJ Quantum Inform. **3**, 2 (2017).

76. Barends, R., Kelly, J., Megrant, A., D. Sank, *et al.*: Coherent Josephson Qubit Suitable for Scalable Quantum Integrated Circuits. Phys. Rev. Lett. **111**, 080502 (2013).

77. Barends, R., Kelly, J., Megrant, A., Veitia A., *et al.*: Superconducting Quantum Circuits at the Surface Code Threshold for Fault Tolerance. Nature **508**, 500 (2014).

78. Blais, A., Gambetta, J., Wallraff, A., Schuster, D. I., *et al.*: Quantum-Information Processing with Circuit Electrodynamics. Phys. Rev. A **75**, 032329 (2007).

79. Upadhyah, R., Thomas, G., Chang, Y.-C., Golubev, D. S., et al.: Robust Strong-Coupling Architecture in Circuit Quantum Electrodynamcis. Phys. Rev. Appl. **16**, 044045 (2021).

80. Majumdar, S., Bera, T., Suresh, R., Singh, V.: A Fast Tunable 3D Transmon Architecture for Superconducting Qubit-Based Hybrid Devices. J. Low Temp. Phys. **207**, 210 (2022).

81. Vepsäläinen, A., Winik, R., Karamlou, A. H., Braümuller, J., *et al.*: Improving Qubit Coherence Using Closed-Loop Feedback. Nature Commun. **13**, 1932 (2022).

82. Werninghaus, M., Egger, D. J., Roy, F., Machnes, S., *et al.*: Leakage Reduction in Fast Superconducting Qubit Gates via Optimal Control. NPJ Quantum Inform. **7**, 14 (2021).

83. Tomonaga, A., Mukai, H., Yoshihara, F., Tsai, J. S.: Quasi-Particle Tunneling and 1/f Charge Noise in Ultrastrongly Coupled Superconducting Qubit and Resonator. Phys. Rev. B 104, 224509 (2021).

84. Marques, J. F., Ali, H., Varbanov, B. M., Finkel, M., et al.: All-Microwave Leakage Reduction Units for Quantum Error Correction with Superconducting Transmon Qubits. Phys. Rev. Lett. **130**, 250602 (2023).

85. Kosen, S., Li, H.-X., Rommel, M., Shiri, D., et al.: Building Blocks of a Flip-Chip Integrated Superconducting Quantum Processor. Quantum Sci. Techn. **7**, 035018 (2022).

86. Ni, Z., Li, S., Zhang, L.., Chu, J., *et al.*: Scalable Method for Eliminating Residual ZZ Interaction between Superconducting Qubits. Phys. Rev. Lett. **129**, 040502 (2022).

87. Mitchell, B. K., Naik, R. K., Morvan, A., Hashim, A., *et al.*: Hardware-Efficient Microwave-Activated Tunable Coupling Between Superconducting Qubits. Phys. Rev. Lett. **127**, 200502 (2021).

88. Kim, M.-D.: Galvanic Phase Coupling of Superconducting Phase Qubits. Appl. Sci. **11**, 11309 (2021).

89. Pita-Vidal, M., Bargerbos, A., Zitko, R., Splitthof, J., *et al.*: Direct Manipulation of a Superconducting Spin Qubit Strongly Coupled to a Transmon Qubit. Nature Phys. **19**, 1110 (2023).

90. Aamir, M. A., Moreno, C. C., Sundelin, S., Biznárová, J., *et al.*: Engineering Symmetry-Selective Couplings of a Superconducting Artificial Molecule to Microwave Waveguides. Phys. Rev. Lett. **129**, 123604 (2022).

91. Maiani, A., Kjaergaard, M., Schrade, C.: Entangling Transmons with Low-Frequency Protected Superconducting Qubits. Phys. Rev. X 3, 030329 (2022).
92. Zhao, P., Lan, D., Xu, P., Xue, G., et al.: Suppression of Static ZZ Interaction in an All-Transmon Quantum Processor. Phys. Rev. Appl. **16**, 024037 (2021).
93. Stehlik, J., Zajac, D. M., Underwood, D. L., Phung, T., et al.: Tunable Coupling Architecture for Fixed Frequency Transmon Superconducting Qubits. Phys. Rev. Lett. **127**, 080505 (2021).
94. Shirai, S., Okubo, Y., Matsuura, K., Osada, A., et al.: All Microwave Manipulation of Superconducting Qubits with a Fixed-Frequency Transmon Coupler. Phys. Rev. Lett. **130**, 260601 (2023).
95. Kubo K., Goto, H.: Fast Parametric Two-Qubit Gate for Highly Detuned Fixed-Frequency Superconducting Qubits Using a Double-Transmon Coupler. Appl. Phys. Lett. **122**, 054001 (2022).
96. Goto, H.: Double-Transmon Coupler: Fast Two-Qubit Gate with No Residual Coupling for Highly Detuned Superconducting Qubits. Phys. Rev. Appl. **18**, 034038 (2022).
97. Moskalenko, I. N., Besedin, I. S., Simakov, I. A., Ustinov, A. V.: Tunable Coupling Scheme for Implementing Two-Qubit Gates on Fluxonium Qubits. Appl. Phys. Lett. **119**, 194001 (2021).
98. Brown, T., Doucet, E., Ristè, D., Ribeill, G., et al.: Trade Off-Free Entanglement in a Superconducting Qutrit-Qubit System. Nature Commun. **13**, 3994 (2022).
99. Wallraff, A., Schuster, D. I., Blais, A., Frunzio, L., et al.: Strong Coupling of a Single Photon to a Superconducting Qubit Using Circuit Quantum Electrodynamics. Nature **431**, 162 (2004).
100. Schuster, D. I., Houck, A. A., Schreier, J. A., Wallraff, A., et al.: Resolving Photon Number States in a Superconducting Circuit. Nature **445**, 515 (2007).
101. Majer, J., Chow, J. M., Gambetta, J. M., Koch, J., et al.: Coupling Superconducting Qubits via a Cavity Bus. Nature **449**, 443 (2007).
102. Hita-Pérez, M., Jaumà, G., Pino, M., Garcia-Ripoll, J. J.: Three-Josephson Junction Flux Qubit Coupling. Appl. Phys. Lett. **119**, 222601 (2021).
103. Sakhouf, H., Daoud, M., Laamara, R. A.: Quantum Process Tomography of the Single Shot Entangling Gate with Superconducting Qubits. J. Phys. B: Atm. Mol. Opt. Phys. **56**, 105501 (2023).
104. Zhou, Yu., Zhang, Z., Yin, Z., Huai, S., et al.: Rapid and Unconditional Parametric Reset Protocol for Tunable Superconducting Qubits. Nature Commun. **2**, 5924 (2021).
105. Córcoles, A. D., Takita, M., Inoue, K., Lekuch, S., et al.: Exploiting Dynamic Quantum Circuits in a Quantum Algorithm with Superconducting Qubits. Phys. Rev. Lett. **127**, 100501 (2021).
106. Takita, M., Cross, A. W., Córcoles, A. D., Chow, J. M., Gambetta, J. A.: Experimental Demonstration of Fault Tolerant State Preparation with Superconducting Qubits. Phys. Rev. Lett. **119**, 180501 (2017).
107. Bravyi, S., Dial, O., Gambetta, J. M., Gil, D., Nazario, Z.: The Future of Quantum Computing with Superconducting Qubits. J. Appl. Phys. **132**, 160902 (2022).
108. Baule, G. M., McFee, R.: Detection of the Magnetic Field of the Heart. Am. Heart J. **55**, 95 (1963).
109. Cohen, D.: Magnetoencephalography: Evidence of Alpha-Rythym Currents. Science 161, 784 (1968).
110. Cohen, D., Edelsack, E. A., Zimmerman, J. E.: Magnetocardiograms Taken Inside a Shielded Room with a Superconducting Point-Contact Magnetometer. Appl. Phys. Lett. **16**, 278 (1970).
111. Zotev, V. S., Matleshov, A. N., Volegov, P. L., Urbaitis, A. V., et al.: SQUID-Based Instrumentation for Ultra-Low Field MRI. Supercond. Sci. Technol. **20**, S367 (2007).

112. Wyss, M., Bagani, K., Jetter, D., Marchiori, E., *et al.*: Magnetic, Thermal, and Topographic Imaging with a Nanometer-scale SQUID-on-Lever Scanning Probe. Phys. Rev. Appl. **17**, 034002 (2022).
113. Szypryt, P., Nakamura, N., Becker, D. T., Bennett, D. A., *et al.*: A Tabletop X-Ray Tomography Instrument for the Nanometer-Scale Imaging: Demonstration of the 1000-Element Transition-Edge Sensor Subarray. IEEE Trans. Appl. Supercond. **33**, 2100705 (2023).
114. Khomchenko, I., Navez, P., Ouerdane, H.: SQUID-Based Interferometric Accelerometer. Appl. Phys. Lett. **121**, 152601 (2022).
115. Vettoliere, A., Granata, C.: Picoammeters Based on Gradiometric Superconducting Quantum Interference Devices. Appl. Sci. **12**, 9030 (2022).
116. Toida, H., Sakai, K., Teshima, T. F., Hori, M., *et al.*: Magnetometry of Neurons Using a Superconducting Qubit. Commun. Phys. **6**, 19 (2023).

Atom Qubits

4

In the Jaynes-Cummings model of Chap. 2, the actual qubit was a two-level system that was created in an atomic-like manner. But, there is nothing to forbid this two-level system to actually lie in the atomic structure of a single atom. In fact, discussion in Chap. 1 pointed out that more than one commercial approach is based upon the concept of trapped ions. The qubit is formed from atomic states of the trapped ion, either the ground and an excited state, or a pair of hyperfine separated states in the ground state itself. And, there is at least one commercial approach which uses neutral atoms as the qubits. In this latter approach, the neutral atoms are trapped on an optical lattice which creates a potential array, which can localize the neutral atom. Again, the qubit can be formed from a variety of the atom's energy levels.

While the development of trapped ion qubits seems to be a newer approach than say Josephson junction qubits, the ideas are probably almost as old [1], and radio frequency traps have been used even longer [2]. The development was rapid with a CNOT gate being demonstrated shortly after the first suggestion, in a structure using a single trapped ion [3]. Technological advances since that time have led to the commercial realization and the expectation of significant future progress [4]. The use of neutral atoms for qubits is somewhat newer, having been developed during the current decade [5, 6]. Often, the neutral atoms are held in place by the use of laser light that creates a potential through a holographic process. In the following, the methods of creating traps for the ion qubit. Then, discussion will turn to the use of these approaches in qubits and qubit circuits. After discussing trapped ion qubits, the discussion will turn to the "traps" for neutral atoms, and the use of neutral atoms for qubits.

© The Author(s), under exclusive license to Springer Nature Switzerland AG 2025   99
D. K. Ferry, *Quantum Information in the Nanoelectronic World*,
Synthesis Lectures on Engineering, Science, and Technology,
https://doi.org/10.1007/978-3-031-62925-9_4

4.1   Ion Traps

Creating ion traps is an application of electromagnetic fields in very complicated physical configurations. The idea is to create a region of space, in which the potential seen by the ion rises in all directions so that it sits in a three dimensional potential well. While there are several types of traps, two distinct types of traps are primarily in use in the experimental world today. The first is the Penning trap, envisioned almost a century ago [7], and in which a magnetic field is used to confine the ion in two dimensions, and a normal quadrupole electrostatic confinement is used in the third direction (along the axis of the magnetic field). The current versions of the trap were developed by Dehmelt, who first experimentally created the ion confinement [8]. An example of such a trap is shown in Fig. 4.1. In this figure, there is a central cylindrical metal tube (opened in the figure so that only the edges in dark blue are visible, with a magnetic field (green arrow) oriented along the central axis. The axial confinement is provided by the two tubes at either end (again in dark blue) which have a voltage bias between them and the central tube. The confinement region is the dark blue ellipsoid of revolution in the center. For this trap, the magnetic field has to be relatively large, so that the cyclotron radius is considerably smaller than the confinement tube. As a result, this type of trap is generally found in various spectroscopic applications. While the Penning trap has been used with qubits, there is a general preference for the Paul type of trap, which provides an entirely electrostatic confinement for the ion [9]. Usually, confinement potentials are used to provide quadrupoles in either two or three dimensions. Hence, two principal versions of the Paul trap exist–the linear Paul trap and a 3D quadrupole Paul trap.

The basic idea of a quadrupole trap is illustrated in Fig. 4.2a [10]. Here, a planar field configuration is shown with four electrodes that produce a potential that varies as one moves around the inner space. Generally, this potential can be an oscillating microwave potential that has the general form

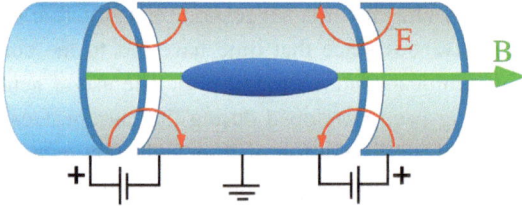

Fig. 4.1 A Penning type ion trap. The axial magnetic field (green) provides confinement in the plane transverse to it, while the voltages, and electric field (red), provide confinement along the central axis. Reprinted from wikimedia commons under the creative commons attribution-share Alike 3.0 license

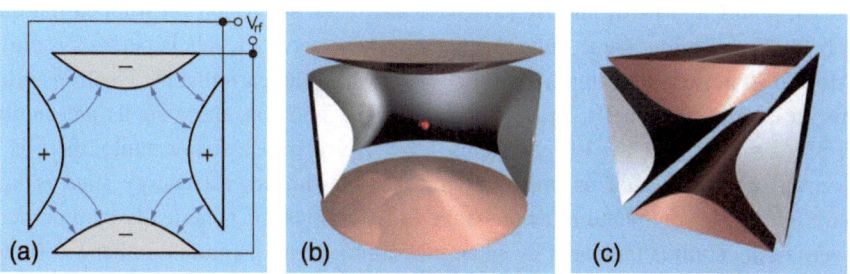

Fig. 4.2 **a** The most basic geometry. The use of hyperbolic shape of the electrodes can produce a perfect quadrupole field. Other configurations are usually just a variation on this theme. **b** The 3D structure, with rotationally symmetric geometry can produce a confining potential in three dimensions. **c** The linear electrodes produce confinement in only two dimensions and must supplemented with an axial confinement potential. Reprinted with permission from [10], copyright 2015 by the American physical society

$$V \sim V_0 + \frac{V_{ac}}{2r_0^2} cos(\omega_t t)(x^2 - y^2) \tag{4.1}$$

in the plane of panel (a). Here, V_0 is a d.c. bias that is often used, V_{ac} is the oscillating voltage that helps to create confinement and control of the trapped ion, ω_t is the frequency of this oscillating voltage, and r_o is the separation between the electrodes. In the plane of this potential, the motion of a trapped particle becomes governed by the Mathieu equation, which is often connected with energy bands (discussed in Chap. 3). Such an energy level is occupied by the trapped particle. The a.c. excitation is typically in the microwave range.

In the 3D quadrupole trap, the above equation is extended to three dimensions, in a structure such as that of Fig. 4.2b. Again, the various electrodes can have both d.c. and a.c. signals. In the linear Paul trap, the quadrupole fields remain in the (x, y)-plane and two end cap electrodes are used to provide a linear, and axial, electric field to help confine the ion. The structure of the linear trap is typically like that of Fig. 4.2c. The electrodes now are often uniform along the length of the trap, but they could be segmented to provide a series traps along the axial direction, and, as will be seen, these can then be used to provide atom motion along the axis of the trap.

4.2 Use of Ion Traps

In one of the earliest suggestions of using trapped ions for qubits, it was suggested that Ba ions could be used [1]. More recently, the choice has changed to Yb ions, in which measurements were originally made on the hyperfine split ground state levels [11]. It was then suggested that trapped ions could be used in a muti-qubit role [12]. It was a few years later that the Yb ion trapping was demonstrated and could be used as a qubit [13].

However, both Ca [14–16] and Be [17] have also been suggested for trapped ion qubits. But, before proceeding, some knowledge about these energy levels is perhaps useful.

Atomic Structure. From high school chemistry, people learn that the electrons orbiting atoms are arranged in shells. Typically, in chemical notation, these shells are numbered 1, 2, 3, ..., with $n = 1$ the lowest energy level. In hydrogen, for example, there is only 1 electron orbiting a nucleus with a charge of 1e. This lowest energy state is known to have no angular momentum, as the wave function must be spherically symmetric, and this zero momentum is indicated by an angular momentum quantum number $l = 0$, which historically is denoted the s-state. Then, the electron in the hydrogen atom is said to reside in the 1s state. Actually, this level can hold two electrons even though hydrogen has only a single electron. Now, as one moves across and down the periodic table, more electrons are added, and gradually more shells become filled. In addition, angular momentum is added in the higher shells. The $n = 2$ shell can hold 8 electrons (4 with each direction of spin). Two of these electrons will reside in the 2s state, so the other 6 must reside in a state with angular momentum, and this set of states is denoted the p states with $l = 1$, and l is denoted as the azimuthal quantum number. Note that there are $2l + 1$ different quantum states, and these are distinguished by a magnetic quantum number $m_z = -l, ..., l$ (-1, 0,1 in the present situation, and the term magnetic implies they will be split in energy by a magnetic field, the Zeeman effect). As one moves up the shells, the angular momentum states historically are denoted by $s, p, d, f, g,$ .... Each shell can hold, 2, 8, 18, and so on. One generally pictures these shells as being thin spherical shells, as the name implies, but this isn't the case. The actual energy levels vary as $-1/n^2$, where $n = n' + l$, and n' is the radial quantum number. Hence, the various angular momentum states have different energies; that is the 2s and 2p states lie at different energies even though they are in the same shell. The situation gets worse, as the d states in the third shell can actually be quite close to the s states of the fourth shell, and they can even overlap for higher shells.

The s and p states consist of 8 electrons that are contributed to each shell. One notices that row 1 of the periodic table has only two elements, H and He, as the electrons for these fill the 1s states. Then the second row has 8 elements that lead to the filling of the 2s and 2p states as one moves from left to right across the table. Hence, for these elements, the position in the table correlates well with the number of electrons in the outer. But, row three continues this behavior with 8 elements filling the 3 s and 3p states. Strangeness now sets in for row four. The problem arises as the 3d states actually lie above the 4s states, but below the 4p states. So, the 4s states are filled in column 1 and 2 on the left side of the table. Then, 10 new columns have appeared in the table, and these correspond to filling the ten 3d states. Finally, on the right side, the six 4p states are filled. This wonkiness continues with later rows, but when the f states have to be filled, one arrives at the lanthanide and actinide series which can't be placed in the nomenclature of the periodic table.

Hence, while Be lies in shell 2 and Ca lies in shell 4, Yb lies in shell 6 as the last of the lanthanides to complete its inner shell filling. While this may all be confusing, it gets worse, because a great many times the notation used is not the standard chemical form, but one familiar in spectroscopy which was used to study optical properties well before the theory of the atom was developed (the full semi-classical–early quantum–theory of even H did not appear until 1916 [18]). This is because the above shell energy structure gets much more complicated. First, the motion of an electron around the atom moves perpendicular to the electric field from the atomic potential, and this (relativistic correction) creates an effective magnetic field that couples to the spin angular momentum; this is the so-called spin–orbit interaction. This is a central part of the fine structure splitting (shown in Fig. 4.3a), which also includes relativistic corrections to the kinetic energy of the electron and contributions arriving from fluctuations in the motion of the electron. The presence of any dipole, such as spin, within the nucleus will also affect the energy levels, due to its electromagnetic coupling to the electrons, and this coupling leads to the hyperfine interaction (shown in Fig. 4.3b).

In spectroscopic notation, the principal quantum number is often omitted and that is the case in this discussion. The spectroscopic notation is in the form of $^{2s+1}X_j$, where s is the total spin quantum number (not to be confused with the s states above) and j is the total angular momentum quantum number. The central X represents the orbital momentum quantum number (l above) now denoted by upper case $S, P, D, F, G, ...$ In

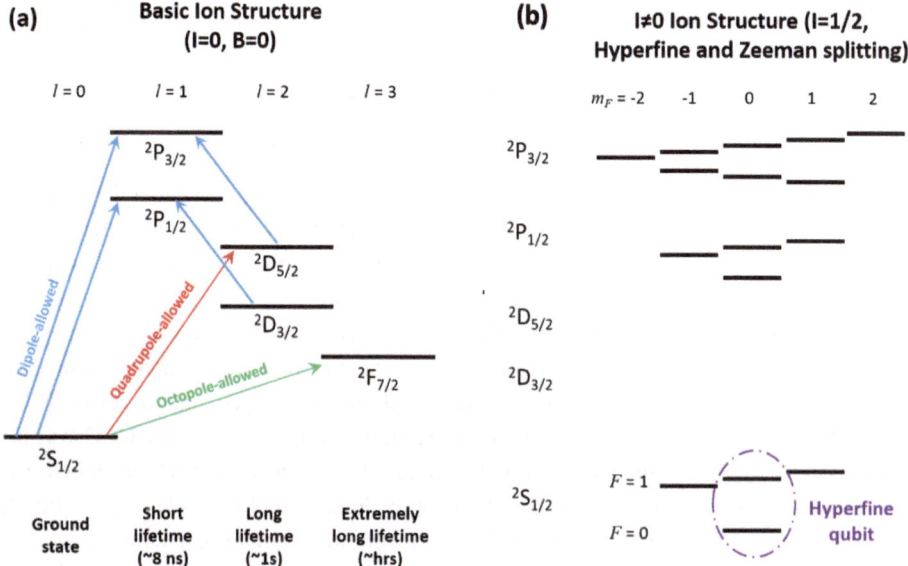

Fig. 4.3 **a** Some basic electronic structure of ions that has been suggested for qubits, when there is no magnetic field and no nuclear spin. The angular momentum states are indicated across the top. **b** Suggested structure for the case in which there is both a small magnetic field and a nuclear spin. The magnetic quantum number m_F (often denoted by m_z) is denoted across the top. Reprinted from [4] with permission from AIP publishing

Fig. 4.3, some example spectroscopic structures and the possible levels that may be used for qubits, are illustrated [4]. Panel (a) denotes a set of energy levels, in the absence of any magnetic field or nuclear spin interaction, that have been suggested for quantum computer qubit applications. Panel (b) shows a set of levels when both a small magnetic field and a non-zero nuclear spin is present. In the latter panel, a nuclear spin of 1/2 has been assumed, and most of the D and F levels have not been shown; here, the ground state hyperfine-split spin levels have been designated as the qubit, and this is often the experimental choice for the qubit. the total angular momentum of the two levels arises from the $\pm 1/2$ of the electron and the $+1/2$ from the nucleus, which gives the two states as 1,0 and one would think should have been labeled as j rather than the value F. The magnetic field splitting arrives from a quite small magnetic field, and these two levels generally have a long lifetime as well as offering a relatively easy set of levels to detect in actual usage. The hyperfine splitting can be typical of the order of 50×10^{-6} eV, which requires quite low temperatures (in the millikelvin range).

Qubits. It was demonstrated fairly early that various levels of the Yb$^+$ ion could be used for a qubit and that the state could be detected [13], with the hyperfine levels indicated in Fig. 4.3b selected. Significant progress toward an actual scalable processor was achieved when it was shown that the ion traps could be effectively created on a silicon surface using standard fabrication techniques [19]. Surface traps on other materials were suggested earlier [20]. This approach has been shown to possess a considerable degree of fault tolerance, which allows the state to be read [21]. The linear Paul ion traps themselves are formed using a combination of 3.07 and 0.27 MHz radio frequencies and d.c. voltage excitation, while the qubits are manipulated using laser excitation [19, 22]. The ions are initially laser cooled to reach their ground state. Quantum gate action is induced by the same laser excitation as for general manipulation.

Coupling between such qubits was demonstrated earlier [23], and entanglement using an optical frequency comb has also been demonstrated [24]. Such entanglement was extended to five qubits quickly [22, 25]. The ability to use error correction techniques with these atomic qubits has also been demonstrated [26].

In many cases, the qubits remain stationary in space, just as is the case for the superconducting transmons in Chap. 3. However, moving charge along a silicon surface is a relatively easy task, with technology similar to either the charge-coupled device (CCD) or a CMOS logic register. The idea of using a technology analogous to the CCD was proposed very early in the ion trap qubit development [27]. The CCD was developed early in microelectronics as a method of moving charge along spatial tracks [28, 29]. The idea is based upon using a series of metal–oxide–Semiconductor (MOS) capacitors to trap and move charge along a surface. The motion is basically a three-step process, although it can be done with two steps. Consider three such MOS capacitors in which the central one is biased with a positive voltage creating an n-type well in normally p-type material. This is similar the normal MOS transistor, except there is no source or drain, so that charge does not flow into the potential well. However, visible light can excite electrons into the potential well and the CCD was a major imaging device in early digital cameras, and still remains important for imaging applications. With charge in the central well, there is no motion of charge, but if the capacitor to the right is now biased to create an n-type well, the charge can begin to move to the right. Lowering the bias of the central well also encourages the charge to move to the right, since the change in potential makes the well to the right a more attractive place for the charge. When transfer is complete, the right-hand capacitor is biased with charge, and the central and left-hand capacitors have no charge in them. Action now moves one capacitor to the right, and the process is repeated. It was recognized that this same action could move ions as well as electrons, and it would be a very efficient process to move the qubits to the gates rather than changing gate properties for stationary qubits [27].

It was realized subsequently that moving the trapped ions via a quantum CCD (QCCD) would make an efficient and scalable quantum computer [30]. This is because the qubit information on one QCCD chip could efficiently be transferred to a second QCCD chip. Say a qubit register exists on each of the two chips, then entanglement between a qubit in one of the registers can be entangled with a qubit in the second register on the second chip. This can be done with a photonic approach so that multiple chips can be used for the processer [30, 31]. In Fig. 4.4 [31], one conceptual form of the photonic link is shown. The qubit information must be used to create entanglement with the photon and then this

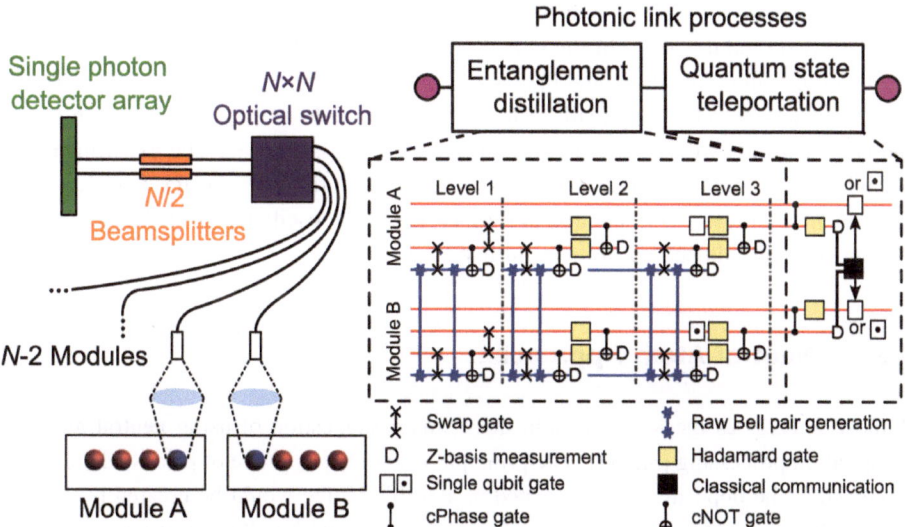

Fig. 4.4 Photonic method of using entanglement to connect two modules (lower left) that have trapped ion qubits. An $N \times N$ optical switch is used to connect the modules, where N is the number of modules. The circuit and diagrams on the right illustrate the optical processing needed for entanglement and teleportation of the qubit information. One can recognize the CNOT and Hadamard gates. Reprinted from [31] under the creative commons attribution 4.0 international license

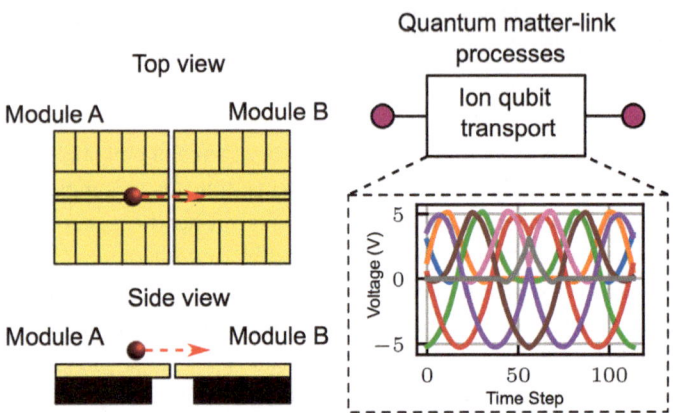

Fig. 4.5 Linking modules (lower left) through direct ion transport is illustrated. The modules are aligned as at the left, and described in the text, and signals are applied between the modules as indicated on the right to establish direct transfer. Reprinted from [31] under the creative commons attribution 4.0 international license

photon is teleported to the new module, using very similar gate schemes as discussed in Sect. 2.5. The various gates used in the entanglement recognition and teleportation are indicated below the circuit diagram in the right-hand panel of the figure.

It is possible that the ion qubit itself actually could be transferred across a small gap between two modules [31]. In the experiment, piezoelectric actuators were used to precisely align both the modules and the movement tracks of the QCCD qubit transfers. This alignment was monitored optically. An electric field could be produced between the two modules, and this would facilitate the ion transfer across the gap. The experiments also established that qubit coherence could be maintained during the transfer. Such a module-module direct transfer is illustrated in Fig. 4.5 [31]. The oscillatory signals applied to the various gates are illustrated in the right-hand panel.

4.3   Neutral Atom Approaches

Almost as early as the ion approach, there was a suggestion of using neutral atoms for information processing [32–34]. The principle is to create a lattice of places where neutral atoms could be deposited (and "trapped"), and this lattice would be created by an array of sites from interfering laser beams. These atoms would be allowed to interact through dipole–dipole interactions. Here, it was felt that decoherence could be suppressed because the neutral atoms would interact with the environment only in a weak manner [32]. Also, operations could be performed on many atoms in parallel. If the qubit states were associated with higher lying (so-called Rydberg) states of the neutral atom, the dipole–dipole interaction could be made larger in the presence of an additional electric field [33]. When an electron is excited from a filled lower level to an empty upper level, a self-energy shift will lower the upper level slightly. When the conditions are right, this level shift will block such transitions in neighboring atoms, an event known as "dipole blockade" or "Rydberg blockade" [34].

But, using laser interference to create the "traps" for the neutral atoms can be difficult. It is known that a laser beam passed through a holographic array will produce the array of traps for the neutral atoms [35]. If this pattern is projected onto a magneto-optical crystal, the resulting array of traps will be long-lived and useful for a quantum processor [36]. Such an approach is shown in Fig. 4.6 [36]. The array of microtraps is generated by a spatial light modulator that imprints a desired spatial pattern onto a single Gaussian laser beam, in this case with a wavelength of 850 nm. This pattern is then projected through a reducing lens onto the magneto-optical material to create the array of microtraps. Here, each microtrap has a radius of about 1 μm, and an energy depth of some 10^{-7} eV. Single atoms in the trap are observed by their fluorescence with an electron-multiplying CCD (EMCCD). A second CCD is used for diagnostics of the trapped array. In this case the system is operated at very low temperature.

A more modern approach using an array of optical confinement defined by the name "optical tweezers." Such a tweezers system is a single optical beam that is highly focused, and this provides a confining force that depends upon the exact nature of the particle and the properties of the surrounding medium. This allows the control and movement of a disordered array of atoms [37]. More recently, the arrays can be produced by using a projected optical signal (produced by diffractive optical elements) that generates a square grid of light lines [38, 39]. Such a line array is shown in Fig. 4.7. The array is produced by combining four laser sources with different frequencies so that a given site has the four

Fig. 4.6 The scheme for generating a microarray of neutral atom traps using holographic imaging. The details are discussed in the text. Reprinted with permission from [36] under the creative commons attribution 3.0 license

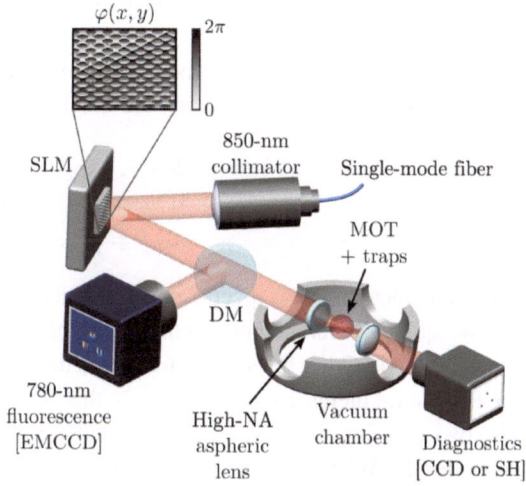

Fig. 4.7 The averaged fluorescene from a 121 qubit array after some signal processing has been done. In the array, performance was determined by using a control (C) qubit which operated on a target qubit (t). Reprinted with permission from [39], copyright 2020 by the American physical society

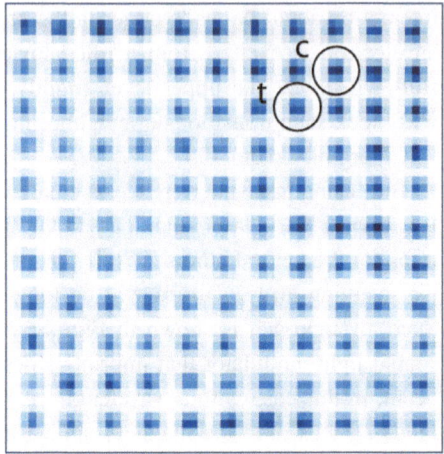

sides defined by lasers that differ by many MHz. This provides cells that are insensitive to phases that arise from variations in the path length of each beam [39]. The fluorescence image in the figure has the background noise reduced by a process known as independent component analysis [38]. In this case, the trapped atoms are Cs atoms that have been optically cooled. One advantage of using such "tweezer" arrays are that individual atoms can be loaded one at a time [40]. This is achieved by improving the algorithm used in setting up the arrays.

4.4   Neutral Atom Qubits

One of the original suggestions for neutral atom qubits proposed using low lying levels of the trapped atom, as these could produce a pair qubits with a single atom. This level scheme (for Cs atoms in this case) is shown in Fig. 4.8. Here, the qubits are formed in the lowest energy shell, $n = 0$, and the magnetic field split spin levels are used. Here, the qubits are defined by [32]

$$|1\rangle_\pm = |F_\uparrow, m_F = \pm 1\rangle$$
$$|0\rangle_\pm = |F_\downarrow, m_F = \mp 1\rangle, \tag{4.2}$$

where the notation is the spectroscopic notation used in Fig. 4.3. The subscript $\pm$ indicates the two qubits. It can be noticed that the qubit lines actually cross as the arrow illustrates for the $+$ qubit. Other lines in the figure are the $\omega_\perp$ and $\omega_\parallel$ are optical trapping frequencies for the atoms themselves, while ω_c is the "catalyst" frequency for operating the gate. The two trapping frequencies are detuned from the blue resonance of the transition to upper levels. When a weak catalyst signal, propagating in the (x, y)-plane, is applied, this provides the excitation for the qubit operation. If this latter signal is tuned to the

Fig. 4.8 Schematic energy levels for an alkali atom in a small magnetic field. The computational basis for polarized light is indicated in the central region with an arrow denoting a qubit transition. The various frequencies are discussed in the text. Reprinted with permission from [32], copyright 1999 by the American physical society

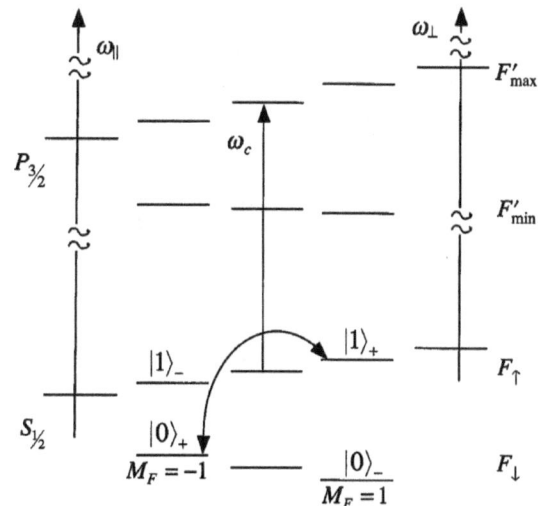

$|S_{1/2}, F_{\uparrow}\rangle \rightarrow |P_{3/2}, F_{max}'\rangle$, as shown in the figure, a dipole is excited if the qubit is in the $|F_{\uparrow}\rangle$, corresponding to the $|1\rangle_{\pm}$, this dipole causes a switch of the $|1\rangle_{-} \otimes |1\rangle_{+}$ combination of the two qubits. Other operations can be obtained by using other forms of this polarized light [32]. As mentioned above, the use of the dipole blockade can block transitions that involve more than a single excitation [34].

The suggestion to use Rydberg states for the neutral atoms seems to have occurred shortly after the previous work [33]. Rydberg states are higher lying states, usually in the outer shells of a multi-electron atom. Some of the lower-lying of these Rydberg states are typically used, while the ground state, or lower qubit level, can be one of the states previously discussed for the qubits. Since these Rydberg states are usually degenerate (many states with the same energy), an electric field is used to split this degeneracy. The scheme is illustrated in Fig. 4.9 [33]. If there are two atoms at fixed positions, the diagonal dipole interactions shift the two levels, while the non-diagonal terms can cause a shift from the state (m, m) to the state $(m \pm 1, m \mp 1)$, where m is the magnetic quantum number for the Rydberg state (see the discussion on atomic structure in Sect. 4.2). The use of Rydberg states means that the energy transition has a much higher value than that of the use of hyperfine levels alone. This can mean a possible higher temperature of operation, providing the atoms can be trapped at these higher temperatures.

Entanglement between two atoms, using Rydberg states, has been demonstrated with relatively long coherence times [41]. In this case, ^{87}Rb atoms were used for the qubits, and a relatively high lying energy level was used for the ground state. This entanglement has been extended over a two-dimensional array of Cs Rydberg atoms [39]. This work has presaged the extension of neutral atoms to an array of various circuits that can be used as a capable processor [42]. Indeed, it has been shown how a neutral atom processor can

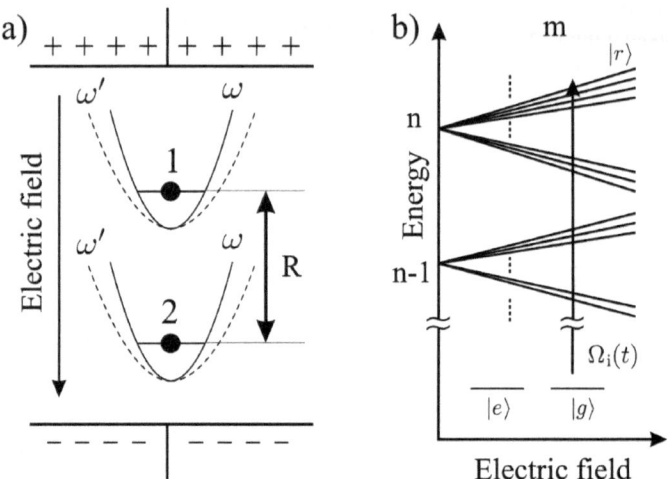

Fig. 4.9 a A constant electric field along the z direction, normal to the plane of atoms, splits the degenerate Rydberg state. **b** Two ground states $|e\rangle$ and $|g\rangle$ and an upper state $|r\rangle$ are indicated as possible qubit states. Reprinted with permission from [33], copyright 2000 by the American physical society

easily do simulation of dense spin systems [43], anti-ferromagnets [6], and other quantum phases of matter [5], all with hundreds of qubits. Even multi-qubit entanglement has been demonstrated [44].

References

1. Cirac, J. I., Zoller, P.: Quantum Computations with Cold Trapped Ions. Phys. Rev. Lett. 74, 4091 (1995).
2. Neuhauser, W., Hohenstatt, M., Toschek, P. E., Dehmelt, H.: Localized Visible Ba+ Mono-Ion Oscillator. Phys. Rev. A 22, 1137 (1980).
3. Monroe, C., Meekhof, D. M., King, B. E., Itano, W. M., Wineland, D. J.: Demonstration of a Fundamental Quantum Logic Gate. Phys. Rev. Lett. 75, 4714 (1995).
4. Bruzewicz, C. D., Chiaverini, J., McConnell, R., Sage, J. M.: Trapped-Ion Quantum Computing: Progress and Challenges. Appl. Phys. Rev. 6, 021314 (2019).
5. Ebadi, S., Wang, T. T., Levine, H., Keesling, A., et al.: Quantum Phases of Matter on a 256-Atom Programmable Quantum Simulator. Nature 595, 227 (2021).
6. Scholl, P., Schuler, M., Williams, H. J., Eberharter, A. A., et al.: Quantum Simulation of 2D Antiferromagnets with Hundreds of Rydberg Atoms. Nature 595, 233 (2021).
7. Penning, F. M.: Die Glimmitladung bei Nedrigem druck zwischen Koaxialen Zylindern in einem Axialem Magnetfeld. Physica 3, 873 (1936).
8. Gebrielse, G., Dehmelt, H., Kells, W.: Observation of a Relativistic, Bistable Hysteresis in the Cyclotron Motion of a Single Electron. Phys. Rev. Lett. 54, 537 (1985).

9. Paul, W., Steinwedel, H.: Notizen: Ein Neues Massenspektrometer ohne Magnetfeld. Zeit. Naturfor. 8, 448 (1953).
10. Brownnutt, M., Kumph, M., Rabi, P., Blatt, R.: Ion Trap Measurements of Electric-Field Noise Near Surfaces. Rev. Mod. Phys. 87, 1419 (2015).
11. Fisk, P. T. H., Sellars, M. J., Lawn, M. A., Coles, C.: Accurate Measurement of the 12.6 GHz "Clock" Transition in Trapped 171Yb+ Ions. IEEE Trans. Ultrason. Ferroelect. Freq. Control 44, 344 (1997).
12. Mølmar, K., Sørensen, A.: Multiparticle Entanglement of Hot Trapped Ions. Phys. Rev. Lett. 82, 1835 (1999).
13. Olmschenk, S., Younge, K. C., Moehring, D. L., Matsukevich, D. N., et al.: Manipulation and Detection of a Trapped Yb+ Hyperfine Qubit. Phys. Rev. A 76, 052314 (2007).
14. Merrill, J. T., Volin, C., Landgren, D., Amini, J. M., et al.: Demonstration of Integrated Microscale Optics in Surface-Electrode Ion Traps. New J. Phys. 13, 103005 (2011).
15. Hrmo, P., Wilhelm, B., Gerster, L., van Mourik, M. W., et al.: Native Qudit Entanglement in a Trapped Ion Quantum Processor. Nature Commun. 14, 2242 (2023).
16. Ballance, C. J., Harty, T. P., Linke, N. M., Sepiol, M. A., Lucas, D. M.: High-Fidelity Quantum Logic Gates Using Trapped Ion Hyperfine Qubits. Phys. Rev. Lett. 117, 060504 (2016).
17. Gaebler, J. P., Tan, T. R., Lin, Y., Wan, Y., et al.: HIgh-Fidelity Universal Gate Set for 9Be+ Ion Qubits. Phys. Rev. Lett. 117, 060505 (2016).
18. Sommerfeld, A.: Zur Quantentheorie der Spektrallinien. Ann. Phys., Ser. 4, 51,1 (1916). (In modern notation, this is vol. 356).
19. Allen, S., Kim, J., Moehring, D. L., Monroe, C. R.: Reconfigurable and Programmable Ion Trap Quantum Computer. Proc. IEEE Conf. Rebooting Comp. (IEEE Press, New York, 2017) 1–3.
20. Kim, T., Maunz, P., Kim, J.: Efficient Collection of Single Photons Emitted from a Trapped Ion into a Single-Mode Fiber for Scalable Quantum-Information Processing. Phys. Rev. A 84, 063423 (2011).
21. Blume-Kohout, R., Gamble, J. K., Nielsen, E., Rudinger, K., et al.: Demonstration of Qubit Operations Below a Rigorous Fault Tolerance Threshold with Gate Set Tomography. Nature Commun. 8, 14485 (2015).
22. Debnath, S., Linke, N. M., Figgatt, C., Landsman, K. A., et al.: Demonstration of a Small Programmable Quantum Computer with Atomic Qubits. Nature 536, 63 (2016).
23. Moehring, D. L., Maunz, P., Olmschenk, S., Younge, K. C., et al.: Entanglement of Single-Atom Quantum Bits at a Distance. Nature 449, 68 (2007).
24. Hayes, D., Matsukevich, D. N., Maunz, P., Hucul, D., et al.: Entanglement of Atomic Qubits Using an Optical Frequency Comb. Phys. Rev. Lett. 104, 140501 (2010).
25. Choi, T., Debnath, S., Manning, T. A., Figgatt, C., et al.: Optimum Quantum Control of Multi-mode Couplings between Trapped Ion Qubits for Scalable Entanglement. Phys. Rev. Lett. 112, 190502 (2014).
26. Mount, E., Kabytayev, C., Crain, S., Harper, R., et al.: Error Compensation of Single Qubit Gates in a Surface-Electrode Ion Trap Using Composite Pulses. Phys. Rev. A 92, 060301 (2015).
27. Kielpinski, D., Monroe, C., Wineland, D. J.: Architecture for a Large-Scale Ion Trap Quantum Computer. Nature 417, 709 (2002).
28. Boyle, W. S., Smith, G. E.: Charge Coupled Semiconductor Devices. Bell. Sys. Tech. J. 49, 587 (1970).
29. Amelio, G. F., Tompsett, M. F., Smith G. E.: Experimental Verification of the Charge Coupled Device Concept. Bell. Sys. Tech. J. 49, 593 (1970).
30. Monroe, C., Kim, J.: Scaling the Ion Trap Quantum Processor. Science 339, 1164 (2013).
31. Akhtar, M., Bonus, F., Lebrun-Gallagher, F. R., Johnson, N. I., et al.: A High-Fidelity Quantum-Matter Link Between Ion-Trap Microchip Modules. Nature Commun. 14, 531 (2023).

32. Brennen, G. K., Caves, C. M., Jessen, P. S., Deutsch, I. H.: Quantum Logic Gates in Optical Lattices. Phys. Rev. Lett. 82, 1060 (1999).
33. Jaksch, D., Cirac, J. I., Zoller, P., Rolston, S. L., et al.: Fast Quantum Gates for Neutral Atoms. Phys. Rev. Lett. 85, 2208 (2000).
34. Lukin, M. D., Fleischhauer, M., Cote, R., Duan, L. M., et al.: Dipole Blockade and Quantum Information Processing in Mesoscopic Atomic Ensembles. Phys. Rev. Lett. 87, 037901 (2001).
35. Newell, R., Sebby, J., Walker, T. G.: Dense Atom Clouds in a Holographic At-om Trap. Optics Lett. 28, 1266 (2003).
36. Nogrette, F., Labuhn, H., Ravets, S., Barredo, D., et al.: Single-Atom Trapping in Holographic 2D Arrays of Microtraps with Arbitrary Geometries. Phys. Rev. X 4, 021034 (2014).
37. Barredo, D., de Léséleuc, S., Lienhard, V., Lahaye, T., Browaeys, A.: An Atom-by-Atom Assembler of Defect-Free Arbitrary Two-Dimensional Atomic Ar-rays. Science 354, 1021 (2016).
38. Lichtman, M. T.: Coherent Operations, Entanglement, and Progress Towards Quantum Search in a Large 2D Array of Neutral Atom Qubits. Ph.D. Thesis, the University of Wisconsin.
39. Graham, T. M., Kwon, M., Grinkemeyer, B., Marra, Z., et al.: Rydberg Mediat-ed Entanglement in a Two-Dimensional Neutral Atom Qubit Array. Phys. Rev. Lett. 123, 230501 (2019).
40. Schymik, K.-N., Lienhard, V., Barredo, D., Scholl, P., et al.: Enhanced Atom-by-Atom Assembly of Arbitrary Tweezer Arrays. Phys. Rev. A 102, 063107 (2020).
41. Levine, H., Keesling, A., Omran, A., Bernian, H., et al.: High-Fidelity Control and Entanglement in Rydberg-Atom Qubits. Phys. Rev. Lett. 121, 123603 (2018).
42. Williams, H. J.: Versatile Neutral Atoms Take on Quantum Circuits. Nature 604, 429 (2022).
43. Browaeys, A., Lahaye, T.: Many-Body Physics with Individually Controlled Ry-dberg Atoms. Nature Phys. 16, 132 (2020).
44. Graham, T. M., Song, Y., Scott, J., Poole, C., et al.: Multi-Qubit Entanglement and Algorithms on a Neutral-Atom Quantum Computer. Nature 604, 457 (2022).

Quantum Processors in Silicon

5

The first idea for a qubit based upon Si arises from the suggestion of Kane to use a ^{31}P dopant[1] in isotopically pure ^{28}Si (silicon, like many materials, has different isotopes that have different numbers of neutrons) [1]. This would enable a silicon platform for quantum technology that would easily integrate with normal silicon microchips. This is an intriguing approach to continue the dominance of silicon in the nanoelectronics, and now the quantum, world. Indeed, some integration of silicon quantum circuits has already appeared [2]. Since the work of Kane, many other qubit concepts, that have their root in the silicon world, have arisen. As a result, other integration platforms have been suggested [3, 4]. These different approaches have naturally led to different suggestions for the implementation of the basic qubit.

Kane's [1] suggestion was to use the nuclear spin of the ^{31}P atom, for which the nuclear-spin quantum number is 1/2, as the qubit, and to use a particular isotope of Si that has no nuclear spin itself. This should then give a relatively low decoherence rate for the qubit [5]. The electron wave function for the positively-charged donors (P is normally an electron-donor impurity used in silicon integrated circuits) is known to extend extensively throughout the conduction band and can mediate the interaction between the nuclear spins on neighboring P atoms. It has also been suggested that actually surrounding the P atom with ^{29}Si, that does have a nuclear spin, would offer the option of other qubits (on the Si atoms) and could yield a multiple qubit register [6]. The use of quadruple donors working together to create a pair of qubits would be another possible option [7].

---

[1] Si has 4 outer shell electrons that bond to neighbors in the tetrahedral coordination of the crystal, and this fills all valence states. The P atom has five electrons and this yields one extra electron that is easily excited to the conduction band and then provides conductivity.

© The Author(s), under exclusive license to Springer Nature Switzerland AG 2025   113
D. K. Ferry, *Quantum Information in the Nanoelectronic World*,
Synthesis Lectures on Engineering, Science, and Technology,
https://doi.org/10.1007/978-3-031-62925-9_5

A different approach uses the electron spin rather than the nuclear spin in a semi-conductor nanostructure [8]. The advantage of this approach lies in the fact that the electron-spin-resonance (ESR) transition for the electron can be effectively tuned by electrostatic gates. Such gates are a common part of the single-electron quantum dot structures discussed in Sect. 3.4 [9], as well as the current world of nanoelectronics is centered on the CMOS structure [9]. In many cases, this is based upon a double quantum-dot structure [2], to be discussed in a following section.

There have been other suggestions for arriving at a meaningful semiconductor qubit, or group of qubits. One approach pointed out that Ge, with a higher conductivity, would allow using a bias-induced electric field shift of the levels [10]. There has also been suggestions for the use of acceptors in place of the donors. There is a feeling that the dopants for the acceptor-based qubit need to be near an interface in order to interact well with surface gates. Hence, the use of acceptors near the surface of both Ge an Si hasbeen studied [11, 12], and entanglement and control have been demonstrated for the acceptor qubits [13].

Nearly all of the discussion above has focused on the use of spin as the dynamic variable in the qubit. But, as was demonstrated in Chap. 3, charge remains a viable participant in qubit operations, and spin-charge hybrid qubits have been discussed [14]. In this chapter, all of these approaches to creating qubits in the silicon system will be treated in the following sections. Each of the various approaches will be described in some detail.

5.1   The Impurity Qubit

As mentioned above, the nuclear spin of the phosphorous atom was suggested by Kane as a suitable basis for a qubit [1]. If this impurity was placed in an isotopically pure ^{28}Si material, which has no nuclear spin, a longer decoherence time could be expected for such a gate [15]. However, one must get the P atoms into the right position. Nevertheless, it is possible to create an atomically precise linear array of single P bearing molecules on a Si surface, and these P atoms can act as quantum qubits.

Getting a P atom to a precise location is actually not so difficult, and there are several methods to achieve this. One approach uses the scanning tunneling microscope (STM) and advanced chemistry to put the P atom in the right position. The first step is to hydrogenate the dangling bonds on the Si (001) surface, which is normally used for nanoelectronics, and then to remove selected hydrogen atoms with an STM (in ultra-high vacuum) [16]. Removing the hydrogen atom leaves a dangling Si bond to which a P-containing molecule may be chemically attached [17]. In one approach, phosphine gas molecules are deposited [18], and with subsequent heating, the phosphine decomposes leaving P attached to the bond. Then, adding another silicon layer buries the P atom in the right position. An alternate approach to placing the P atom is to pursue single atom implantation [19–22]. This technology was developed some years ago, and a more modern application even used an STM to "implant" the atoms [23]. Both approaches have been used to fabricate P-atom qubits.

Prior to putting the P atoms into place, the problem is to obtain a suitable Si layer in which to embed the P atom. Of course, this can be done by growing an entire boule of the material from initially purified silane (SiH_4) to create the polycrystalline material from which the isotopically pure boule is grown [24], but this may well be overkill. Naturally-occurring silicon contains approximately 92% ^{28}Si, so that removing the remainder is a refinement problem initially. It also is possible to actually grow just a thin layer of the material with the desired isotopically pure state, by refining silane, and using vapor-phase epitaxy [25]. A different approach is to ion implant an entire layer of ^{28}Si in an aluminum exchange process [26]. In the latter situation, aluminum is first deposited on the Si surface, and it is this layer into which the proper Si isotope is implanted. Then, upon annealing, the implanted layer will migrate to the interface between the aluminum and the silicon substrate. Following this the metal can be removed.

Once the P atom has been embedded within the desired silicon layer, it is necessary to find and control its properties. Detection of the nuclear spin has been achieved with Hall devices [27], by creating a single-electron transistor in the silicon in which P has been embedded [28], or by a noninvasive spatial metrology technique based upon measuring the various properties of the atom [29]. Finding the donor allows one to measure the transport through the region [30], and single atom memories have been created as a result [31]. Of course, the nuclear spin qubit has also been measured [32].

Progress in the P atom qubit has advanced to create pairs of qubits by inserting two P atoms quite close together [33]. In this case, the qubit is connected to charge moving between the quantum wells created by the two P atoms. Such a qubit will be described below with the creation of similar quantum wells by electrostatic gates. Similarly, double quantum dots have been prepared by using multiple P atoms via STM lithography [34–36]. Such a double dot, embedded within a single electron structure is shown in Fig. 5.1 [36]. Spin blockade, along with normal Coulomb blockade is observed in these qubits. This has led to the development of a single-triplet qubit architecture that is thought to be scalable [37]. A schematic of how the double donor can be measured with electron spin resonance (ESR) is given in Fig. 5.2 [38]. In these qubits, there is both the nuclear spin and the electron spin that contribute to the overall system, as indicated in the upper right panel. In the lower right panel, the ESR signal is depicted. The arrows beside the ESR lines indicate the direction of the nuclear spins on the two P atoms. From studies such as this, it is clear that the qubit may be easily addressed through the ESR signal [39].

It is clear from the above that the spin of the electron can be as important as the nuclear spin, and that both may be utilized. As a result, it has been suggested to use the electron spin as a separate qubit state [40, 41]. If the P atom is close to the Si/SiO_2 interface, then the electron can be drawn to the surface by a gate, and shared between the donor and a surface state. This separates it somewhat from the donor itself, and provides a relatively long-range coupling between the two qubits. Rotations of the two qubits can be achieved

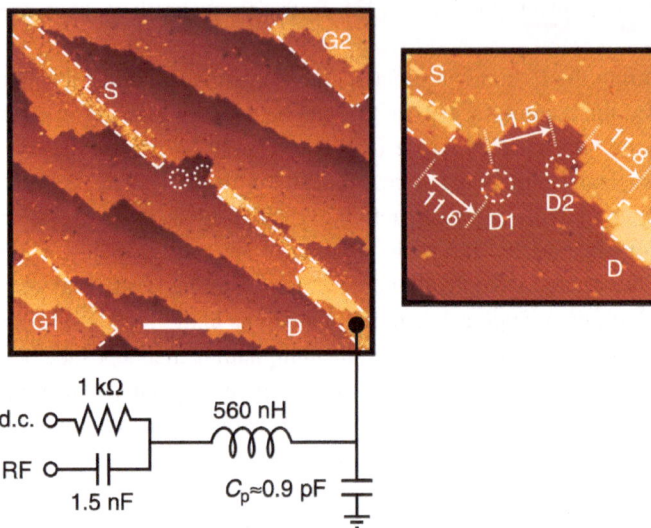

Fig. 5.1 An STM image of a double quantum dot, indicated by D1 and D2 in the blowup at the right. These dots are composed of two and three P atoms, respectively. The positions of metal electrodes are indicated also by dashed lines. The source (S) and drain (D) electrodes are for current to pass, as indicated in the bias circuit shown at lower left. Reprinted from [36] under the Creative Commons Attribution 4.0 License

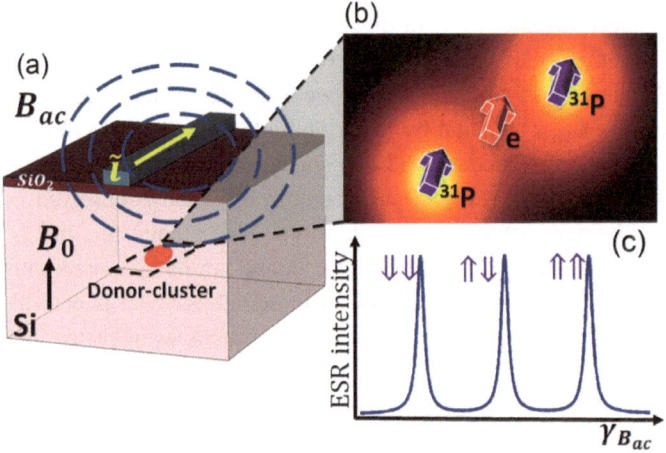

Fig. 5.2 **a** A schematic of a donor dot qubit in silicon with ESR measurements via a magnetic field, both d.c. and a.c. **b** The potential landscape around the two P atoms (red is lower into the potential well). Here, there are two P atoms and a single bound electron. **c** The schematic of the ESR measurement. Reprinted from [38] under the Creative Commons Attribution 4.0 License

with the application of an electric field, and the resulting pair of qubits has been named the flip-flop qubit [41, 42]. In this situation, the placement of the P atoms is not so critical. The actions of the flip-flop qubit are also controlled by an additional microwave signal [42, 43]. It has also been established that parallel gate operations can be achieved with high fidelity with these gates [44]. Such a flip-flop qubit is shown schematically in Fig. 5.3 [44]. Here, it may be seen that the electron is drawn to the interface, while the actual donor remains at its desired location. Embedding this qubit within an access and control circuit is illustrated in Fig. 5.4 [45]. The electron and atom spins are schematically shown at the lower right of panel (a) which is a micrograph of the actual device. The grey lines are the metal forming the quantum dot which resides within a single-electron transistor (SET, shown in the figure). The qubit is addressed via both dc bias and a microwave signal. Panel (b) shows the energy levels of the two qubits. The thin and thick arrows represent the electron and nuclear spins, while the red and blue arrows are the microwave transitions. A magnetic field of ~ 1.5 T is applied, and the microwave signals are given approximately as 42 GHz ± 97 MHz.

Besides P, there have been suggestions to use other atoms in Si as qubits. These include the donors S and Se and the acceptor Zn [46], which form deep levels and can lead to other defect states that may be useful. In some sense, these deep levels have similarities to the quantum dot qubit to be discussed below, as the deep level can form such a quantum well for the electron or hole. The use of a heavy-hole state around an acceptor has also been suggested [47], as has the use of SiC [48] or Ge [49] as the substrate, rather than Si. It has been demonstrated that the qubit state of these impurity-related qubits can be read with precision [50, 51].

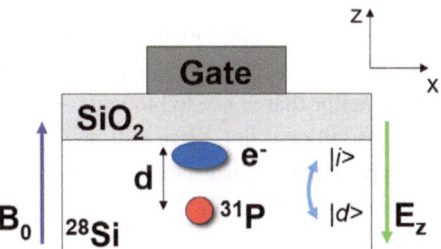

Fig. 5.3 A schematic of the flip-flop qubit. A donor atom of ^{31}P is embedded in the isotopically pure ^{28}Si at some distance (d) from the interface. Applying an electric field via the metal gate (on top) pulls the electron to the surface, so that it is no longer located on the impurity. A constant magnetic field is applied to create the spin states, with the two qubits noted as donor $|d\rangle$ and interface $|i\rangle$. Reprinted from [44] under the Creative Commons Attribution License

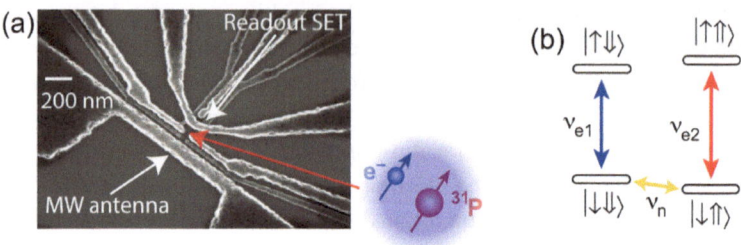

Fig. 5.4 **a** A scanning electron micrograph of the single P atom device. The red arrow denotes the position of the P atom, with the two spins shown schematically at the lower right. **b** The qubit energy levels with the spin denoted by the arrows in the states and the transition frequencies indicated by the red and blue arrows. Reprinted with permission from [45], copyright 2018 by the American Physical Society

5.2   The Silicon Double Quantum Dot

One of the earliest qubit descriptions used the electron spin rather than the nuclear spin in a double quantum dot defined by electostatic gates entirely fabricated on a semiconductor surface [8]. One strength of this approach is said to lay in the fact that the ESR transition for the electrons in these dots can be effectively tuned by the electrostatic gates used to create the dots. In a sense, this quantum dot structure is an extension of the single-electron dots discussed in Sect. 3.4 [9]. But, the double dot system is somewhat more complicated. It is necessary to consider how a pair of dots may be coupled to each other in a simple manner, and thus can be viewed as an "artificial dot molecule". As is the case with real molecules, in which molecular orbitals are formed as a result of the interaction between the wave functions of the component atoms, here the overlap gives rise to new electronic states that can be understood as molecular in nature. In Sect. 3.4, the conditions for which the Coulomb blockade existed was found by determining the total energy of the circuit. Then, this energy was minimized with respect to the tunneling of a single electron through one of the two capacitors (Fig. 3.7). This provided a charging diagram in which tunneling could occur at the crossing points of a set of diagonal lines. With double quantum dots, there is now an additional capacitor that is needed to make the two dots, which are shown in black in Fig. 5.5. With the additional dot, there will now also occur an additional gate voltage and non-tunneling capacitor to adjust the charge on the second dot. In the figure, the charging parts of the circuit are shown in red, and the overall bias voltage is shown in green.

A classical analysis of the energetics of Coulomb-coupled quantum dots has been performed by several authors, in which the confinement-induced quantization of energy within the dots is ignored [52–54]. The details of this analysis will not be repeated here, but the end result is a complicated behavior. In Fig. 5.6, the results of this are shown in

Fig. 5.5 The circuit representation for a double single-electron quantum dot structure. The two dots are shown as the black circles, while the charge adjusting gates are shown in red. The applied bias voltage source is the green circle. This can be compared to the single dot structure of Fig. 3.7

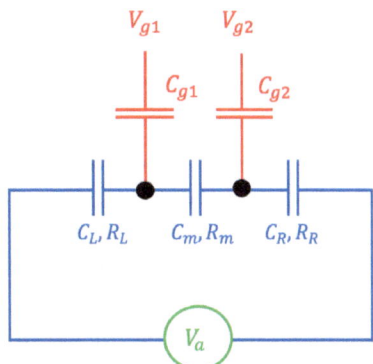

a plane defined by the two gate voltages of Fig. 5.5. Here, the charging lines appear vertically and horizontally in panel (a) in which no interaction between the dots is assumed. The number of electrons in dot 1 (the left dot in Fig. 5.5) and dot 2 (the right dot in Fig. 5.5) are indicated by the notation (n_1, n_2). One can see how the position in the charging plane changes as the amount of charge on one of the dots is incremented (or decremented) by a single charge.

When the dots interact with each other, the contour becomes distorted, and the original four-fold intersections of the squares that result from the crossing of the charging lines in panel (a) of Fig. 5.6 develop into a pair of nearby triple points. One such pair of triple

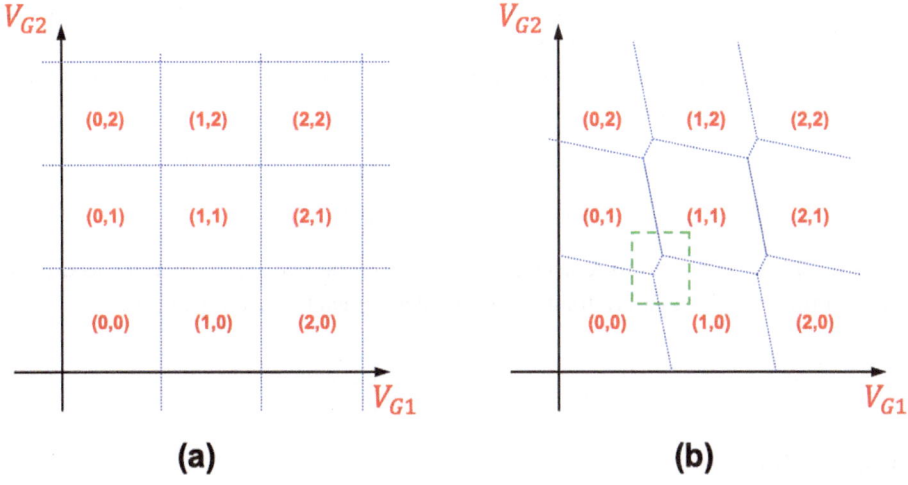

(a) **(b)**

Fig. 5.6 The charging diagram for the double-dot system of Fig. 5.5. **a** The charging lines for when the two dots do not interact. **b** The resulting charging diagram when the two dots interact. The four-fold crossing of panel (a) now diverge into two closely spaced triple points, and current may flow between. The electron population in the two dots is shown in each panel as (n_1, n_2)

points is outlined in green in panel (b). The existence of these triple points is critical
for transport, as they allow for current flow even when the source–drain bias (V_a) is
small. The various voltages allow the number of electrons on the two dots to change by
a single electron on either of the dots. Passing between the two triple points signal the
passage of a single electron between the dots, changing the number on each of the dots
(one increasing, one decreasing). These triple points arise from the mutual capacitance
between the two dots. When this capacitance is non-zero, the charging of one of the dots
by an additional single electron modifies the electrostatic energy of the other, which is
why both populations change.

To understand how important this connecting line, between the two triple points
becomes, we expand upon the Schrodinger equation of (A.2) and write it as

$$-\frac{\hbar^2}{2m}\frac{\partial^2\psi}{\partial x^2} + \frac{1}{2}m\omega^2x^2\psi \equiv H\psi = E\psi, \tag{5.1}$$

in which H is the Hamiltonian operator (representing all the differential terms on the left,
the first of which is the kinetic energy, and the second is the potential energy). Now,
assume that the two dots have energy levels E_1 and E_2 so that we may expand (5.1) into
an eigenvalue expression for the two critical energy levels in the dots, for the case in
which there is no interaction between the two dots. These equations then become

$$H\,|1\rangle = E_1\,|1\rangle$$
$$H\,|2\rangle = E_2\,|2\rangle. \tag{5.2}$$

It is important to understand that the two energy levels will move as the biases V_{g1}
and V_{g2} are varied. This, in fact, is reflected by the various lines in Fig. 5.6. At some
point, the two dots will interact, and this will provide an interaction energy $\Delta/2$. When
this happens, (5.2) becomes

$$E_1|1\rangle + \tfrac{\Delta}{2}|2\rangle = E|1\rangle$$
$$E_2|2\rangle + \tfrac{\Delta}{2}|1\rangle = E|2\rangle, \tag{5.3}$$

where (5.2) has been used to replace H. In this case, the two initial energies lead to two
new (hybrid) energies that are found from the determinant of the Eqs. (5.3):

$$\begin{vmatrix} (E_1 - E) & \tfrac{\Delta}{2} \\ \tfrac{\Delta}{2} & (E_2 - E) \end{vmatrix} = 0, \tag{5.4}$$

and this leads to

$$E = \frac{1}{2}(E_1 + E_2) \pm \frac{1}{2}\sqrt{\varepsilon^2 + \Delta^2}, \tag{5.5}$$

where the detuning is defined as $\varepsilon = E_1 - E_2$. This now produces two hybrid energies. When the two dot energies are equal, the detuning is zero, and this corresponds to the line between the two triple points (in the green box) of Fig. 5.6b. In many applications, the upper hybrid frequency is used as the new $|1'\rangle$ state of the qubit and the lower hybrid is then taken as the $|0'\rangle$ of the qubit. This pattern will be repeated in many of the discussions below. At resonance, the two hybrid states are separated by Δ.

The energy structure, and the existence of the triple points is a general property of any double quantum dot system in which tunneling coupling and the Coulomb blockade are present. This pair of triple points is shown in Fig. 5.7 for the double phosphorous quantum wells of Fig. 5.1 [36]. One might recall from Sect. 4.2, that the atomic energy levels had angular momentum, just as the quantum dots may have. There, the lowest state was the s level, which had only a single wave function and no angular momentum; it may be referred to as the singlet state. The p level had three angular momentum states (the so-called magnetic quantum numbers), and so is often called a triplet state as these degenerate levels are split by a magnetic field. This same terminology is usually adopted for the quantum dot states. Figure 5.7 is an experimentally determined view of the part of Fig. 5.6b that lies with the green box. In this latter figure, ε is the energy difference between the (0,4) and (1,3) states, while $\Delta\varphi$ is the phase shift for the transition between the two states. Panel (a) arises when no magnetic field is present, and the state (1,3) is a single degenerate level. Panel (b) illustrates the change that occurs when a magnetic field is present and the (1,3) splits into the triplet state (three levels). In the latter case, there is very little phase shift in the transition.

Fig. 5.7 The stability diagram for the double quantum dot structure from two implanted P quantum wells of Fig. 5.1. **a** The diagram when no magnetic field is present. The quantity ε is the detuning between the two states, while $\Delta\phi$ is the phase shift. **b** The connection between the triple points is removed in a maagnetic field. Reprinted from [36] under the Creative Commons Attribution 4.0 License

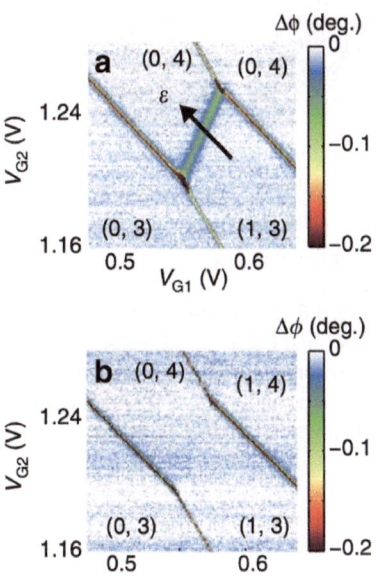

The most interesting to some is the lithographically fabricated silicon double-dot system. Such an electrostatic double-dot system is shown in Fig. 5.8 [55]. The structure, shown in panel (a) and panel (c), is fabricated on isotopically pure ^{28}Si. The confinement gates TC, MC, and BC (in purple) create the quantum dots electrostatically, illustrated in panel (c). The gate LP (red) separates the two dots and creates the tunneling barrier between them. The gate LB is used to apply voltage pulses to the dots for readout purposes. The gates marked S create the single-electron transistor that is used to feed charge into the two dots, as indicated in panel (c). The micromagnet creates a local magnetic field which also assists in reading out the states of the dots. The charging diagram in panel (b) shows the two triple points and the states (0,1) and (1,0), that appear on either end of the "detuning" arrow. The detuning here is the reverse of the quantity ε of Fig. 5.7, and the phase change of this latter figure is now the conductance dI_S/dV_{LP}, where I_S is the current through the single-electron transistor and V_{LP} is one of the aforementioned gate voltages. The charging diagram of panel (b) is plotted using the voltages on probes BC and LP, shown in panel (a).

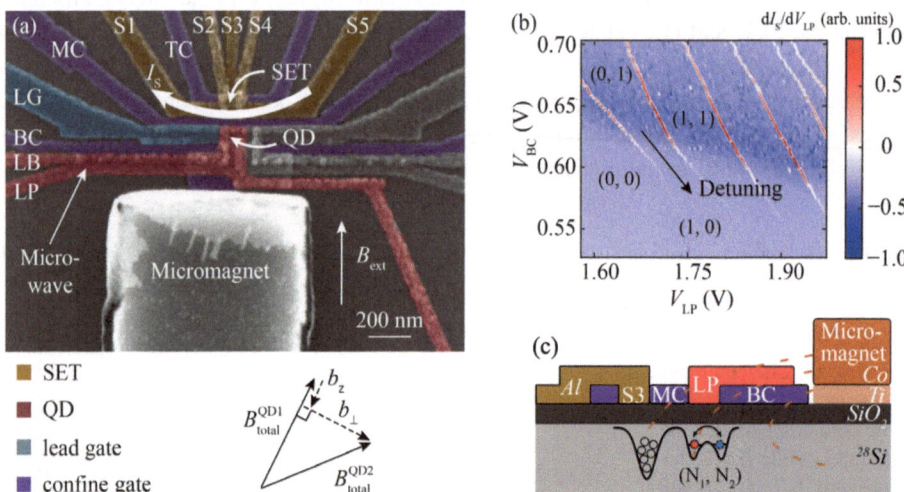

Fig. 5.8 An electrostatically defined spin qubit. **a** A false color image of the device. The various lines and colors are discussed in the text. **b** The charging diagram of the double-quantum dots. **c** A schematic of the various layers in the structure, and the formation of the single-electron transistor (left) and the two dots indicated by the number of electrons in each. Reprinted from [55] with permission of AIP Publishing

5.3   Spin Qubits

In the original suggestion for electrostatically designed quantum dots [8], it was suggested that the spin of an excess electron localized in a single-electron quantum dot would be a viable qubit. Then, two such dots created a pair of qubits with the interaction between mediated by a tunnel barrier. The latter could be controlled by the bias applied to the metallization that creates the barrier. Such tunneling gate controlled interactions between a pair of quantum dots had been demonstrated earlier [56]. When the barrier is high, there is no interaction between the two dots. However, when the barrier is lowered, perhaps via a pulsed voltage, electrons in the two dots could be considered to interact through the exchange interaction. Since, this initial work, there has been considerable effort in the area of such spin qubits.

Spin qubits which formed electrostatically-defined quantum dots in the Si/SiO2 system have typically had from 1 to 3 electrons in the dots [57]. Interaction is somewhat different than originally proposed as a pulsed microwave ESR signal is used to exercise coherent control over the qubit(s). Generally, in such Si dots, a problem arises with level degeneracy.

Normally, in the inversion layer of an MOS structure in Si, the six ellipsoids of the conduction band break into a two-fold and a four-fold set due to the inversion potential leading to quantization at the Si/SiO_2 interface. The four fold set arises from the effective mass normal to the interface in these four valleys being small, while it is the larger heavy mass that is normal to the interface in the two-fold set of valleys (the quantization energy is inversely proportional to the mass [9]). It is this two-fold pair of equivalent valleys that provide the actual states used in the qubit. However, because of the degeneracy of these two valleys, the lowest are doubly degenerate from this valley degeneracy and also double degenerate from the spin degeneracy (thus, becoming four-fold degenerate). What is needed is what is termed valley splitting, in order to remove the former degeneracy [58]. The role of disorder on this valley splitting has also been studied [59]. In the above qubit system [57], valley splitting was obtained and its effect upon the qubit operations measured.

A typical gate defined double quantum dot system is illustrated in Fig. 5.9 [60]. In panel (a), a false color electron micrograph of the device is shown. The two small blue circles are the double dot system; the large blue circle to the left is a sensor dot. The various voltages are labeled. The gold squares are ohmic contacts to the device through which the signals are grounded, except for the one on the far left through which the sensor dot can feed information about the state of the dots. A small Co micromagnet is placed on top of the device to induce a magnetic field around the dots. Panel (b) illustrates the operation of each dot. Here, during initialization the spin down state is created as the only state into an electron can tunnel. Next, the gate voltages assure that the electron state is pushed deep into the Coulomb blockade region after which a microwave signal is applied to gate C to induce dipole ESR. Finally, the gate voltages are reduced to the readout level,

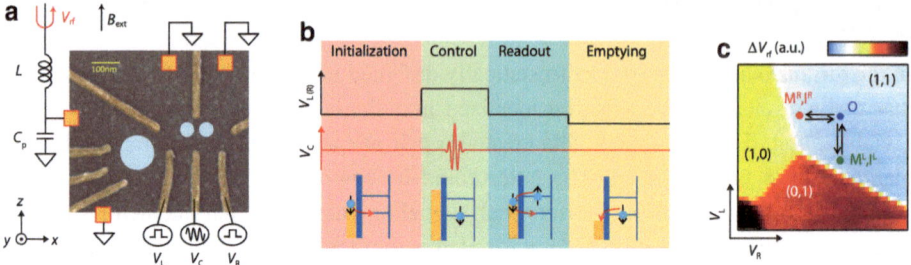

Fig. 5.9   **a** A false color electron micrograph of the device (details are discussed in the text). **b** Application of various voltages to manipulate the state of the dots. **c** Charging diagram for the double dot system. Reprinted from [60] under the creative commons attribution 4.0 license

so that only the spin up electron can tunnel out. The final stage merely empties the dot (or resets the dot) for the next operation. The charging diagram, corresponding to Fig. 5.6b appears in panel (c).

Another prototypical double quantum dot spin qubit gate that is created by the use of surface gates is shown in Fig. 5.10 [61]. Two spin qubits are created in the double dot system that forms an artificial molecule. Each of the two quantum dots has its own robust shell structure and is filled with a number of electrons. The higher electron number in the dots helps to screen static electric fields that may arise from disorder in the crystal structure, and this provides greater stability to the qubits. Panel (a) of the figure shows a visualization of the quantum dot structure, while panel (b) is a blowup of the dot region, with the potential landscape illustrated below it. This latter is a scan along the yellow line extending from gate RG, in which this line is the axis of the potential and dot structure shown in the lower part of (b). Gates G1 and G2 correspond to the dot bias in Fig. 5.5, and serve to adjust the charge in each dot. On the left of Fig. 5.10 is the single-electron transistor (SET) that is used to monitor the dot-dot interaction, the latter of which affects the conductance through the SET. Panel (c) shows the charging diagram in the absence of a magnetic field, where the expected structure of Fig. 5.6b is most evident on the left- and right-hand ends of this panel. The numbers in parentheses represent the two dot occupations, as usual. The color scale (panel (c)) may be seen to correspond to the current through the SET. The lower scale is the detuning defined as $\varepsilon = V_{G1} - V_{G2}$. Gate J and voltage V_J control the barrier between the two dots, whereas the detuning affects the potential minima difference between the two dots. Panel (d) shows the shift in ESR frequency of the two dots Q1 and Q2, while the color scale indicates the adiabatic inversion population (in terms of the electron charge). The individual dots are charged from the reservoir of electrons that exist under gate G (where there is an inversion layer in the p-type substrate). The micromagnet (the Co layer) gives a gradient in the magnetic field, when it is activated, and this is used to perform a set of single gate operations in the two quantum dots, and the interdot potential V_J is used to control the two-dot gate interactions

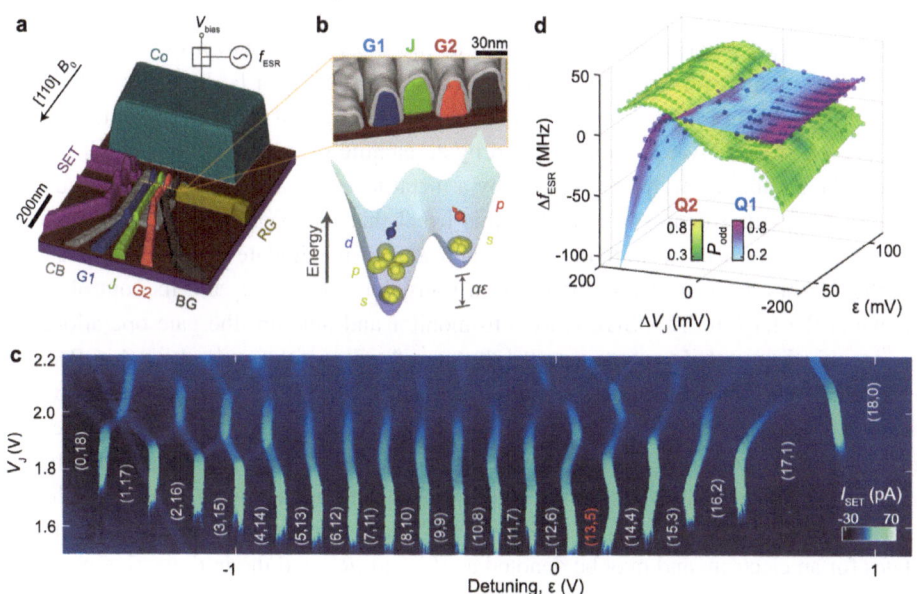

Fig. 5.10 The device overview of a two spin qubit system. **a** A schematic visualization of the device. The details are discussed in the text. **b** Creation of the double dots with the gates (top) and the potential wells formed by the gates (bottom). **c** Charging diagram for the two dots. **d** Resonance frequency of the two dots when a microwave with frequency $30.486 + \Delta f_{ESR}$GHz is applied. Here, $\Delta V_J = V_J - 1.58$V. Reprinted from [61] under the creative commons attribution 4.0 license

via the exchange coupling between the spins of the highest lying electrons in each dot. The authors demonstrated the capability for Bell state preparation and tomography of the states.

Various other structures using double quantum dots have been used to construct swap gates [62, 63]. A set of gates has also been used to "shuttle" the qubits in a manner similar to a charge-coupled device and the ion shuttle described in Chap. 4 [64]. Finally, a method of addressing all qubits (not just one or two) at the same time with a global microwave field has been developed [65]. Others have suggested using Si/SiGe for spin qubits [64, 66, 67], or holes in Si [68].

5.4   Charge Qubits

While the preceding section focused on using the spin as the dynamic variable for the qubit, some early studies of silicon quantum dots focused instead upon the charge itself. Such charge qubits have a relatively long coherence time [69], but are often thought to be overly sensitive to fluctuations and noise [70, 71]. One such relatively early charge qubit is shown in Fig. 5.11 [72]. Here, the qubit is created on a silicon-on-insulator (SOI) layer; the buried oxide is often referred to as the BOX layer, shown in panel (b) of the figure. The qubit is formed with the top phosphorous-doped Si layer that is only 35 nm thick. The qubit is formed by electron-beam lithography that patterns the active region, and the top silicon layer is then removed by an etching process. The grey areas in panel (a) are the remaining Si, mostly topped by a metal layer for access. The double quantum well is the qubit structure indicated in this panel that has been isolated from the other Si. Six gates are used to initialized the qubit and to manipulate its operation. There is another quantum dot in the single-electron transistor (SET) line, and its control gate is shown at the far left. This SET is used to monitor and measure the gate operation, as it is capacitively coupled to the qubit and thus can "see" the charge in the qubit. Panel (c) illustrates the double quantum well and the charge levels. The external gates control the amount of charge through the electrostatic coupling to the isolated double dot. The qubit is defined by an electron in the upper-most energy level of the double-dot structure. It should be noted that electrons cannot enter or leave the double dot because of its physical isolation from any of the gates or remaining Si. The two quantum wells have well-defined states for an electron, and may be denoted as $|L\rangle$ and $|R\rangle$, and the initialization makes the left state occupied. Then, when a short voltage pulse (top central gate in the left panel of the figure), of 5 ns duration, is applied, the two states are brought together to create any desired superposition, and this remains after the pulse is removed. This gate operation is shown in Fig. 5.12. The quantities Δ and ε are the energy of interaction between the two dots and the detuning $\varepsilon = E_L - E_R$, respectively. The solid blue lines are the two hybrid energies that arise when the dots interact.

Other charge qubits have been created with one [73, 74] and two qubits [75] in the same system. In some of these cases, a microwave signal was used to provide coupling between the two qubits in the system. Quantum charge oscillations have been observed both in the single qubit [76] and in coupled qubits [77], establishing that gate operation can be obtained in such charge qubits. Operations have also included valley variables [78]. More recently, the CNOT gate has been demonstrated [79].

Quite often, single-electron devices are still used to inject charge into the qubit structure, as has been shown in a chain of these qubits [80]. The use of microwaves to read out the state of the qubit has also been used [81, 82]. Finally, it should be noted that these qubits have also been formed using holes, rather than electrons [83].

5.5 Hybrid Qubits

Of course, as in previous discussions of various operational regions of qubits, there can arise qubits which are somewhere between spin and charge, known here as hybrid qubits. In fact, it has suggested that the the hybrid qubits in the MOS structure would actually allow a higher density of quantum information than either of the pure qubits discussed

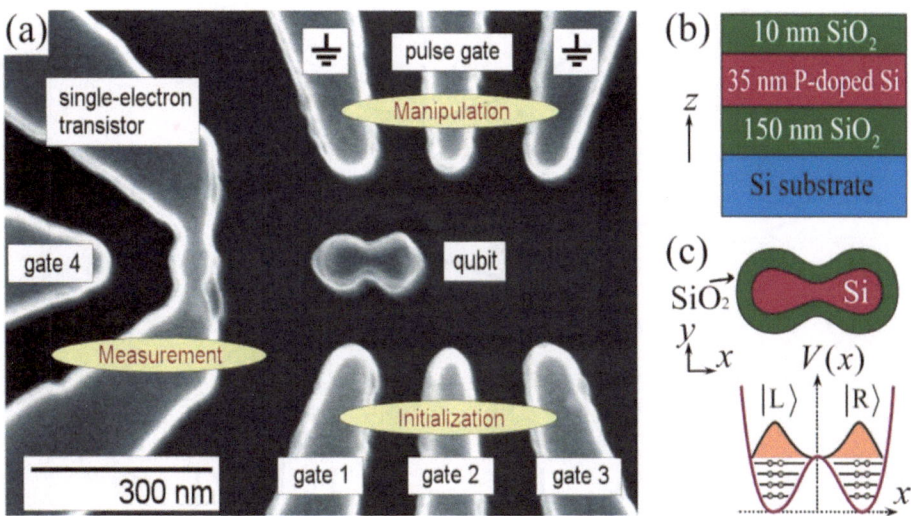

Fig. 5.11 a A scanning electron micrograph of a Si charge qubit structure. The grey areas are remaining Si, which are used to apply potentials to the qubit in the center. **b** The material structure before lithography and etching. **c** The double quantum dot (top) and potential structure of the double quantum well (bottom). Reprinted with permission from [72], copyright 2005 by the American Physical Society

above [84]. A hybrid qubit arises when the dynamical variable depends upon both spin and charge. The hybrid qubit requires neither nuclear-state preparation nor micromagnets for control, and becomes considerably more amenable to systems [14, 85]. The hybrid qubit uses a traditional double dot system, but three electrons are used in the system, with the gates tuned so that there are two electrons in the one dot and one in the other dot to define the (1,2) state, or the reverse situation for the (2,1). The structure and these two states, along with their interaction diagram are shown in Fig. 5.13 [86]. Such a three-electron occupation of the two dots is sometimes called the singlet–triplet qubit [87]. In the figure, panel (a) shows a micrograph of the device with the gates labeled (grey areas). Gates L, M, R define the dots and the barrier between them. Gates LP and RP adjust the occupancy of the two dots. Gate T operates with gate M to control the dot coupling. The structures marked as QL and QR are used to measure the occupancy of the two dots.

Operation of the qubit of Fig. 5.13 is shown in Fig. 5.14, and involves a microwave signal being applied at the gate R [86, 88]. The dot is initialized in the configuration (1,2). At this point the energy levels in the two dots are equal. Then, a voltage is applied to the left dot, and a microwave signal is applied to the right dot to push one of the electrons up to the triplet level (upper right). This state can then be read by this excited electron tunneling out to the right contact (lower right). Following this, the qubit can be reset/initialized back to the first situation. The singlet (lower energy level with both dots in this

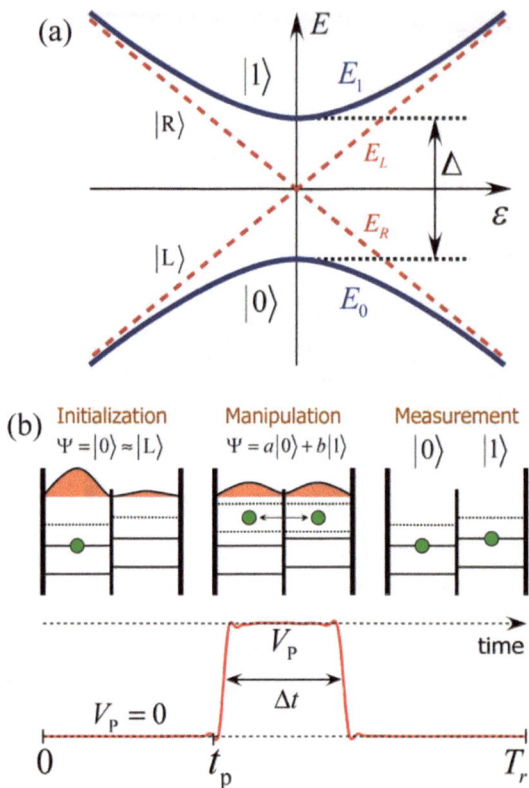

Fig. 5.12 The process of initialization, manipulation and measurement for a Si charge qubit. **a** The energy diagram for the hybridization of the energies of the left and right quantum wells. The two hybrid energies represent the two qubit states. **b** The qubit is initialized in the left quantum dot, then manipulated with a pulsed voltage applied to the pulse gate (shown in Fig. 5.11) for $\Delta t = 5ns$ (shown below). Measurement is made with the SET shown Fig. 5.11. Reprinted with permission from [72], copyright 2005 by the American Physical Society

level) state in the right dot, state (1,2), is taken to be the $|0\rangle$ state. The triplet state in the right dot, the (1,2) state, with the electron excited to the triplet state, is taken to be the logical $|1\rangle$. The presence of the extra electron in the right dot means that the hybrid states of the dots are not pure singlet or triplet and fast electric field techniques can be used to manipulate the qubit in either X or Z rotations on the Bloch sphere.

Using three electrons in the double quantum dot extends the coherence of the qubit [89]. More recently, the importance of the spin–orbit interaction in these gate-defined dots has been studied [90, 91]. Quantum dots have also been created in GaAs [92]. Coherence times in the tens to hundreds of microseconds have been found for these hybrid qubits [93], and there are expectations that this will lead to viable quantum processors [94].

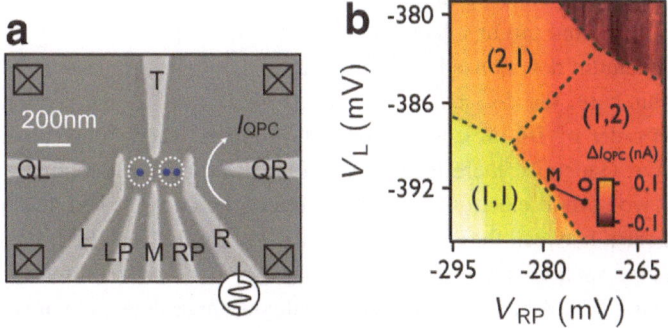

Fig. 5.13 A microwave driven hybrid qubit in a Si double dot device. **a** An SEM image of the device. The light grey areas are the metal gates. **b** The charging diagram for the qubit operation. Reprinted from [86] under the creative commons attribution 4.0 license

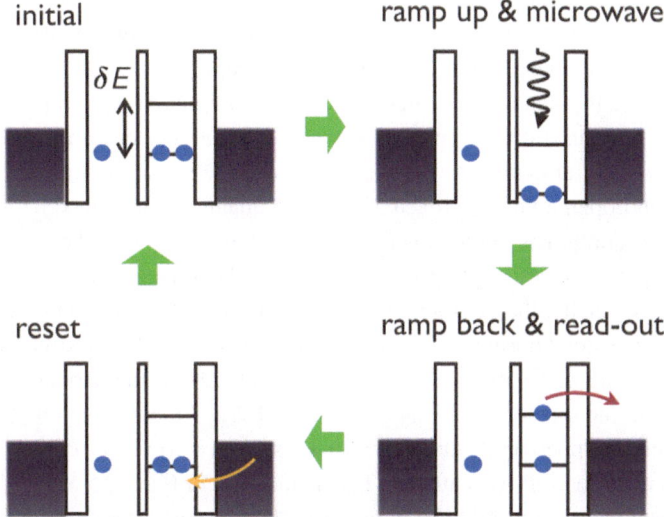

Fig. 5.14 Operation of the qubit of Fig. 5.13. Initialization is shown in the upper left in the (1,2) state with the two energy levels degenerate. Resulting operations are explained in the text. Reprinted from [86] under the creative commons attribution 4.0 license

5.6 Silicon Surface Bond Qubits

A relatively new approach to a silicon qubit has appeared that is based upon a surface dangling bond. Earlier in this chapter, surface processing of Si was discussed in relation to the incorporation of single P atoms. One of the methods discussed there was the hydrogenation of the Si (001) surface and then removing single H atoms with an STM. In the surface bond approach, the dangling bond that remains after removing the H is used. This bond has two states that can be adapted to serve as a qubit [95]. The dangling bond plays the role of a quantum dot, as used in the various silicon qubits discussed in the previous sections. In this case, the dots can be separated by as little as the spacing of adjacent atoms (on the order of 1/4 nm), although slightly larger spacings are typically used, as discussed below. Generally, there is an electron localized on this dangling bond, and this particle can have several electron states. If electron pairs are created at adjacent atoms, the tunneling rates between these two quantum wells can be very high [96]. As demonstration, a line of such quantum dots was created so that the electrons were coupled in an "atomic" wire [97]. As the host material is silicon, these dots and wires can be gated. This means that the occupation of the bond states at each quantum dot can be varied with external gates [98].

As with other quantum dot qubits, single charges can be controllably moved between the two wells, and actually can be shuttled along a chain of paired quantum wells. One can now consider how these chains can be used to create qubit gates useful in quantum computing [99]. The charge state can be gated so that its energy level may be moved above or below the bulk Fermi level (the energy midway between that of the highest occupied state and the next higher state that is unoccupied). As with other silicon gating structures, this local potential movement may be a few hundred meV, which suggests that this process may be useful at room temperature. Nearly all current embodiments of quantum computers work at extremely low temperatures. The possibility of room temperature operation should open a new approach to creating quantum computers.

Creating the qubits has been shown to be relatively easy with the proper hydrogen-based lithography. The Si dangling bonds on the (001) surface consist of two bonds per surface atom. Normally, the H atom attaches to only one of these atoms, while the other hybridizes with a neighboring Si atom, and this leads to a "reconstruction" of the Si surface: this pairing of Si atoms leads to a new unit cell on the surface that is a factor of two larger in one direction. This is the so-called 2×1 reconstructions (as one moves from lower left to upper right, the atomic lines show a periodicity of two atoms closer together with a space every two atoms). This is pictured in Fig. 5.15 as an STM image of the hydrogenated surface [100]. The dimension of the reconstructed cell is shown by the yellow bars. The position of a qubit is indicated by the yellow box and green dot (where a H atom has been removed). In qubit operation, this position is taken as the logical zero.

While it may seem that designing qubits and their circuits from these atomically-defined dangling-bond qubits will be difficult, a design tool has been created to aid in the task [101]. This multi-level tool is expected to address most of the problems in such system design. The overall development of the technology for quantum computation is being pursued by Quantum Silicon, Inc., a Canadian company that hopes to join those companies listed in Chap. 1.

Fig. 5.15 An STM image of a reconstructed Si (001) surface. The yellow lines indicated the two cell distances of the new surface cell. The yellow box indicates a single qubit, in which the H atom has been removed from the area of the green dot. Reprinted from [100] under the creative commons attribution 4.0 license

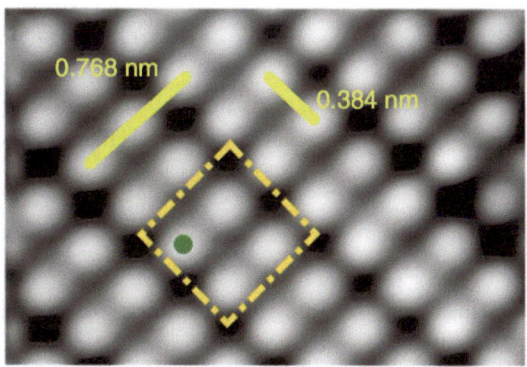

5.7 Silicon Quantum Sensors

Silicon, and other semiconductors sensors are perhaps the most common sensors in the world today. Infrared optical detectors, using the optical transition between an impurity level and the conduction or valence band, have been around since the properties of silicon were first studied. The most common solar energy device today is the Si solar cell. And, Si MOS transistors along with GaAs high-electron mobility transistors (HEMTs) dominate the microwave first stage amplifier world, sensing microwave signals. At low temperature, these transistors are universally used in radio astronomy because of their low noise capabilities.

Sensing magnetic fields is accomplished for a great many applications with the Hall effect sensor. The Hall effect was discovered in the late nineteenth century [102] and is used as a standard measurement tool in the semiconductor world. When a current flows, for example in the z-direction, along a semiconductor of finite width, and a magnetic field is exerted in the x-direction (normal to the conducting plane of the semiconductor), a resulting force is directed in the y-direction. This force is termed the Lorentz force, but results from Maxwell's equations of electromagnetics. If the sample width (in the y-direction) is not infinite, charge piles up at the edges of the sample and creates a voltage in this direction. Generally, one writes the voltage as

$$V_y = -RI_z B_x . \tag{5.6}$$

Here, R is a material constant that depends upon the local carrier(s) charge and density. Along with materials characterization, Hall effect sensors have been used in most magnetic storage devices for decades (although they are gradually being replaced with more sensitive techniques, such as the TMR sensors discussed in Chap. 3).

Roughly a century after Hall's original work, it was found that, at low temperatures, (5.6) was no longer obeyed. One can define the Hall resistance R_H as

$$R_H = \frac{V_y}{I_z} = -RB_x \, . \tag{5.7}$$

From (5.6), one expects the Hall resistance to increase linearly with B_x. At low temperature, it was discovered that the Hall resistance did not increase linearly, but that it increased by steps with values of [103]

$$R_H = \frac{h}{e^2 v}, \, v = 1, 2, 3, \dots. \tag{5.8}$$

This result became important first for determining the value of the fine structure constant[2] and earned von Klitzing the Nobel prize in 1985. In 2018, the von Klitzing resistance $R_K = h/e^2$ became the international standard of resistance, and the entire set of SI standards was reworked to reflect this new fundamental constant whose value is 25812.80745... ohms. One should note the relation of this constant to the quantum unit of flux $\Phi_0 = h/e$, which is now $2.067833848 \times 10^{-15}$ Weber. Thus, the Quantum Hall Effect (QHE) becomes a measurement for currents, flux, and voltage, although it only standardizes the ratio of voltage to current. Nevertheless, the QHE is a significant force in quantum measurements, and therefore in quantum sensing.

In Chap. 3, the SQUID was discusssed and the role quantized flux played in its operation was pointed out. This is not the only sensor used to measure quantized flux. In semiconductors and thin metallic films, one can make a small loop or ring, in which the Aharonov-Bohm effect [104] is measured. When such a ring, with contacts on the opposite sides (so that the current entering or leaving through the two contacts travels around one or the other of the two sides of the ring), has a current passing through the two contacts, and a magnetic field passing through the ring (normal to its plane), its resistance will oscillate with the value of this magnetic field with a period of

$$\frac{eBA}{\hbar} = 2\pi v, \, v = \frac{\Phi}{\Phi_0} \, . \tag{5.9}$$

Here, A is the area of the ring, B is the magnetic field passing through the ring ($\Phi = BA$), and $\Phi_0 = h/e$. Placing such a ring on the cantilever of an STM creates a powerful scanning magnetometer to measure local magnetic fields with near atomic resolution.

Cantilevers, such as those whose tips create an STM, or an atomic force microscope (AFM), or the above scanning magnetic microscope, can easily be made using standard microchip processing. Thus, these actuators, or sensors can be integrated with the electronics needed to process the signal and give the desired digital output. For example, a cantilever with a metallic dot placed on the tip, and a metallic or heavily-doped layer

---

[2] The fine structure constant is $(2\epsilon_0 c R_K)^{-1}$, where c is the speed of light in vacuum and ϵ_0 is the permittivity of free space. Since ϵ_0 and c are fundamental constants in their own right, the fine structure constant is most accurately determined by the value of R_K. The fine structure levels of an atom were discussed in Chap. 4.

in the semiconductor below the cantilever yields a capacitance measurement of the cantilever deflection. This can be used to detect temperature variations, or many-times the acceleration whose force bends the cantilever. In fact, this is the basis of the accelerometers in many auto applications, particularly in air bags. In the AFM, the deflection of the tip over the surface that is being measured is detected via optical interferometry or by using the piezoelectric effect within the cantilever (in which a bending force induces an electric field that can be measured).

Since qubits are quite sensitive to their environment, they have become a basis for new approaches in quantum sensing. Generally, these qubits respond to electric and magnetic fields, and to temperature [105], so any physical quantity that can be measured via these three quantities, or in which the variable to be measured can be converted into one of these three quantities, is subject to sensing with qubits. This of course includes the measurement of charge states, as these are easily converted to an electric field or voltage. In essence, using qubits to measure or sense a variable relies upon the ability to read the qubit state in a reliable manner. In this regard, the nature of measurement in strongly interacting spin systems has been described [106]. In Fig. 5.16, a qubit used to read the spin state is shown [107]. In panel (a), the usual double quantum dot and single-electron transistor used to sense the charge are shown. Across the top is a microwave strip-line used to apply a microwave field. Gates CB, G2, and G4 define the two quantum dots, while gates G1 and G3 are the control gates for dot occupation. The dots are filled from the reservoir gate RG. The experiment starts with a single electron in dot 2, which is the (0,1) state. Then, a second electron is loaded into dot 2, and the qubit initialized by transferring one of the electrons to dot 1, the (0,2) → (1,1) transition, as shown in panel (b). This transferred electron is then spin polarized by the ESR signal into either the spin up or spin down as desired, while the (1,1) state is sensed by the SET. Measurement is determined by a singlet–triplet transition, in which the electron can only transfer back to dot 2 if its spin is opposite to that of the electron remaining in dot 2 (spin blockade). Whether or not the electron has transferred can be determined by the time required to pump another electron into the double-dot system, as the transition from (1,1) to (1,2) is quite fast, while the opposite case is slow. The spin blockade creates the temporal separation of the two possible pathways. Similar spin transitions have also been used in a triple dot system to achieve high fidelity readout of the spin state [108]. In some cases, error correction can be applied to the pulse width needed to separate the two states mentioned above [109].

In both Chap. 3 and here, the single-electron tunneling transistor (SET) has been discussed with both a single dot (for example, in the SQUID of Chap. 4) and double dots (as discussed in Sect. 5.2 above). This is because nearby charge, such as may be found in a nearby qubit, converts to a gate voltage on the SET, mimicking the action of the voltage source V_g in Fig. 3.7, or the two gate voltages in Fig. 5.5. In a sense, the capacitive coupling between the quantum dot of the SET and one of the quantum dots of the qubit serves to convert charge in the qubit to a gate voltage seen by the SET. This voltage is simply $V = Q/C$, where Q is the charge in the qubit and C is the capacitance between

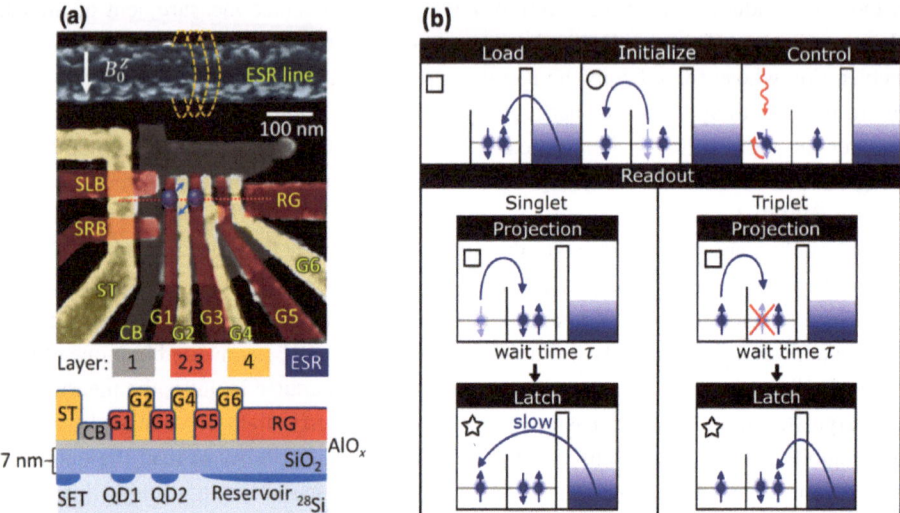

Fig. 5.16 **a** A false-color micrograph of a qubit used as a sensor. There is a two quantum dot qubit (in purple along the dashed red line) and a SET in the gold circuit on the left that is used as a readout of the qubit. **b** Operation of the qubit (top) and the various readout approaches, discussed in the text. Reprinted from [107] under the creative commons attribution 4.0 license

the two quantum dots. Since the capacitance varies as the inverse of the distance between the quantum dots, the closest dot to the SET dominates the signal read by it. Thus, the SET becomes an excellent charge sensor for any qubit, and it may be seen in most of the above figures of qubits in this chapter. The dependence on distance is key to the operation of the sensor in Fig. 5.16.

Charge sensing has been done with single quantum dot SETs in many applications [110–113]. Others have used radio frequency signals to determine the properties of the SET, which is itself used as the charge sensor [114–117]. In addition, the SET has been expanded with the use of a double quantum dot structure to enable higher temperature operation [118].

New forms of magnetic sensing have begun to rely upon the appearance of devices that exhibit tunneling magnetoresistance (TMR). The TMR device is composed of two magnetic films, separated by a thin tunneling barrier, typically an oxide. One of the films is thick enough to not be affected by an external magnetic field (the *fixed* layer), while the second is thin enough to have a free magnetization that can rotate in the presence of an external magnetic field (the *free* layer). The principle is that in the quiescent state (no external magnetic field), the polarizations of the two films is parallel, and electrons can tunnel through the barrier while retaining their spin polarization. However, when an external magnetic field is applied, the thin layer tries to align its polarization with the external field, and this reduces the tunnel current exponentially, so that the resistance of

Fig. 5.17 a Outline of the TMR chip and the magnetic field concentrators (MFC), which are high permeability materials in which the external magnetic field lines are pulled into these layers, thus increasing the magnetic field that is sensed. **b** A blowup of the TMR/MFC region. **d** The TMR device, in which the fixed polarization material is on the and the free layer is on the left (the free responds to the external magnetic field and raises the tunnel resistance). Reprinted from [119] under the creative commons attribution 4.0 license

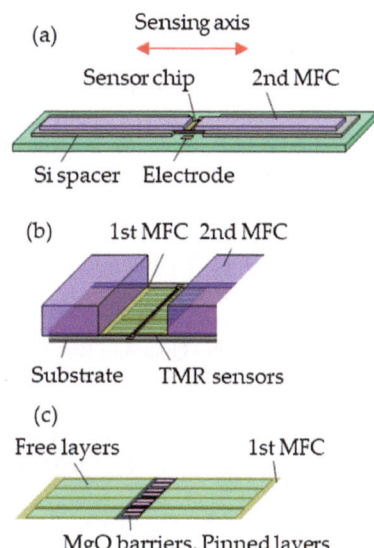

the device increases exponentially. In Fig. 5.17, such a TMR device and its environment are illustrated [119]. The external magnetic field is concentrated and aligned using high permeability materials by the MFC layers in Panel (a). Panel (b) shows a blowup of the region where the MFC layers meet the TMR device. (c) The TMR device with the free layer on the left-hand side of the tunneling region (in purple). Such devices can be used at higher temperatures than SQUIDs and are usually cheaper to manufacture. As the latter paper points out, the sensitivity of the TMR device, with the MFC layers, is enough to avoid having to screen out stray magnetic fields with a magnetically shielded environment (the special rooms in which MRI measurements are carried out). The TMR devices have found many applications in using biological magnetic fields for medical analysis [120, 121].

References

1. Kane, B. E.: A Silicon-Based Nuclear Spin Quantum Computer, Nature **393**, 133 (1998).
2. Holman, N., Rosenberg, D., Yost, D., Yoder, J. L., *et al.*: 3D Integration and Measurement of a Semiconductor Double Quantum Dot with a High Impedance TiN Resonator. NPJ Quantum Inform. **7**, 137 (2020).
3. Thomas, C., Michel, J.-P., Deschaseaux, E., Charbonnier, A., et al.: Superconducting Routing Platform for Large-Scale Integration of Quantum Technologies. Mater. Quantum Techn. **2**, 035001 (2022).
4. De Michielis, M., Ferraro, E., Prati, E., Hutin, L., *et al.*: Silicon Spin Qubits from Laboratory to Industry. J. Phys. D: Appl. Phys. **56**, 363001 (2023).

5. O'Brien, J. L., Schofield, S. R., Simmons, M. Y., Clark, R. G., *et al.*: Towards the Fabrication of Phosphorus Qubits for a Silicon Quantum Computer, Phys. Rev. B **64**, 161401 (2001).

6. Wolfowicz, G., Mortemousque, P.-A., Guichard, R., Simmons, S., *et al.*: ^{29}Si Nuclear Spins as a Resource for Donor Spin Qubits in Silicon. New J. Phys. **18**, 023021 (2016).

7. Pakkiam, P., Houe, M. G., Koch, M., Simmons M. Y.: Characterization of a Scalable Donor-Based Singlet-Triplet Qubit Architecture in Silicon," Nano Lett. **18**, 4081 (2018).

8. Loss, D., DiVencenzo, D.: Quantum Computation with Quantum Dots. Phys. Rev. A **57**, 120 (1998).

9. Ferry, D. K.: Transport in Semiconductor Mesoscopic Devices. (IOP Publishing, Bristol, UK, 2015) Chapter 8.

10. Sigillito, A. J., Tyryshkin, A. M., Beeman, J. W., Haller, E. E., *et al.*: Stark Tuning of Donor Electron Spin Quantum Bits in Germanium," Phys. Rev. B 94, 125201 (2016).

11. Abadillo-Uriel, J. C., Calderón M. J.: Interface Effects on Acceptor Qubits in Silicon and Germanium," Nanotechn. **27**, 024003 (2016).

12. Salfi, J., Tong, M., Rogge, S., Culcer, D.: Quantum Computing with Acceptor Spins in Silicon," Nanotechnol. **27**, 244001 (2016).

13. Abadillo-Uriel, J. C., Salfi, J., Hu, X., Rogge, S., *et al.*: Entanglement Control and Magic Angles for Acceptor Qubits in Si. Appl. Phys. Lett. **113**, 012102 (2018).

14. Kim, D., Shi, Z., Simmons, C. B., Ward, D. R., *et al.*: Quantum Control and Process Tomography of a Semiconductor Quantum Dot Hybrid Qubit. Nature **511** 70 (2014).

15. Morello, A., Pla, J. J., Bertet, P., Jamieson, D. N.: Donor Spins in Silicon for Quantum Technologies. Adv. Quantum Technol. **3**, 200005 (2021).

16. Lyding, J. W., Albein, G. C., Shen, T.-C., Wang, C., Tucker, J. R.: Nanometer Scale Patterning and Oxidation of Silicon Surfaces with an Ultrahigh Vacuum Scanning Tunneling Microscope. J. Vac. Sci. Technol. B **12**, 3735 (1994).

17. Hersam, M. C., Guisinger, N. P., Lyding, J. W.: Silicon-Based Molecular NanoTechnology, Nanotechnol. **11,** 70 (2000).

18. Simmons, M. Y., Schofield, S. R., O'Brien, J. L, Curson, N. J., *et al.*: Towards the Atomic Scale Fabrication of a Silicon-Based Atomic-Scale Quantum Computer. Surf. Sci. **532-535**, 1209 (2003).

19. Matsukawa, T., Fukai, T., Suzuki, S., K. Hara, et al.: Development of Single-Ion Implantation—Controllability of Implanted Ion Number. Appl. Surf. Sci. **117/118**, 677 (1997).

20. Matsukawa, T., Shinada, T., Fukai, T., Ohdomari, I.: Key Technologies of a Focus Ion Beam System for Single Ion Implantation. J. Vac. Sci. Technol. B **16**, 2479 (1998).

21. Jamieson, D. N., Yang, C., Hearne, S. M., Pakes, C. I., *et al.*: Controlled Single-Ion Implantation in Silicon Using an Active Substrate for Sub-20keV Ions. Appl Phys. Lett. **86**, 202101 (2005).

22. Mitic, M., Andresen, S. E., Yang, C., Hopf, T., *et al.*: Single Atoms Si Nanoelectronics Using Controlled Single-Ion Implantation. Microelectron. Engr. **78-79**, 279 (2005).

23. Liu, G., Shao, X., Chen, C., Wang, X., *et al.*: Controlled Implantation of Phosphorous Atoms into a Silicon Surface Lattice with a Scanning Tunneling Microscopy Tip. ACS Appl. Electron. Mater. **3**, 3338 (2021).

24. Becker, P., Pohl, H.-J., Riemann, H., Abrosimov, N. V.: Enrichment of Silicon for a Better Kilogram. Phys. Stat. Sol. A **207**, 49 (2010).

25. Mazzocchi, V., Sennikov, P. G., Bulanov, A. V., Churbanov, M. F., *et al.*: 99,992% ^{28}Si CVD-Grown Epilayer on 300 mm Substrates for Large Scale Integration of Silicon Spin Qubits. J. Cryst. Growth 509, **1** (2019).

26. Schneider, E., England, J.: Isotopically Enriched Layers for Quantum Computers Formed by ^{28}Si Implantation and Layer Exchange. ACS Appl. Mater. Interfaces **15**, 21609 (2023).

27. Sleiter, D., Kim, N. Y., Nozawa, K., Ladd, T. D., et al.: Quantum Hall Charge Sensor for Single-Donor Spin Detection in Silicon. New J. Phys. **12**, 093028 (2020).
28. Mahapatra, S., Büch, H., Simmons, M. Y.: Charge Sensing of Precisely Positioned P Donors in Si. Nano Lett. **11**, 4376 (2011).
29. Mohiyaddin, F. A., Rahman, R., Kalra, R., Klimeck, G., et al.: Noninvasive Spatial Metrology of Single-Atom Devices. Nano Lett. **13**, 1903 (2013).
30. Miwa, J. A., Mol, J. A., Slafi, J., Rogge, S., Simmons, M. Y.: Transport Through a Single Donor in p-Type Silicon. Appl. Phys. Lett. 103, 043106 (2013).
31. Freer, S., Simmons, S., Laucht, A., Muhonen, J. T., et al.: A Single-Atom Quantum Memory in Silicon. Quantum Sci. Technol. **2**, 015009 (2017).
32. Muhonen, J. T., Laucht, R., Simmons, S., Dehollain, J. P., et al.: Quantifying the Gate Fidelity of Single-Atom Spin Qubits by Randomized Benchmarking. J. Phys.: Cond. Matt. **27**, 154205 (2015).
33. Hudson, F. E., Ferguson, A. J., Escott, C. C., Yang, C., et al.: Gate-Controlled Charge Transfer in Si:P Double Quantum Dots. Nanotechn. 19, 195402 (2008).
34. Weber, B., Mahapatra, S., Watson, T. F., Simmons, M. Y.: Engineering Independent Electrostatic Control of Atomic Scale (~4nm) Silicon Double Quantum Dots. Nano Lett. **12**, 4001 (2012).
35. Weber, B., Tan, Y. H. M., Mahapatra, S., Watson, T. F., et al.: Spin Blockade and Exchange in Coulomb Confined Silicon Double Quantum Dots. Nature Nanotechn. **9**, 430 (2014).
36. House, M. G., Kobayashi, T., Weber, B., Hile, S. J., et al.: Radio Frequency Measurements of Tunnel Couplings and Singlet-Triplet Spin States in Si:P Quantum Dots. Nature Commun. **6**, 8848 (2015).
37. Pakkiam, P., House, M. G., Koch, M., Simmons, M. Y.: Characterization of a Scalable Donor-Based Singlet-Triplet Qubit Architecture in Silicon. Nano Lett. **18**, 4081 (2018).
38. Wang, Y., Chen, C.-Y., Klimeck, G., Simmons, M. Y., Rahman, R.: Characterizing Si:P Quantum Dot Qubits with Spin Resonance Techniques. Sci. Repts. **6**, 31830 (2015).
39. Hile, S. J., Fricke, L., House, M. G., Peretz, E., et al.: Addressable Spin Resonance Using Donors and Donor Molecules in Silicon. Sci. Adv. **4**, eaaq1459 (2018).
40. Pla, J. J., Tan, K. Y., Dehollain, J. P., Lim, W. H., et al.: High Fidelity Readout and Control of a Nuclear Spin Qubit in Silicon. Nature **496**, 334 (2013).
41. Tosi, G., Mohiyaddin, F. A., Schmitt, V., Tenberg, S., et al.: Silicon Quantum Processor with Robuts Long-Distance Qubit Couplings. Nature Commun. **8**, 450 (2017).
42. Calderon-Vargas, F. A., Barnes, E., Economou, S. E.: Fast High-Fidelity Single-Qubit Gates for Flip-Flop Qubits in Silicon. Phys. Rev. B **106**, 156302 (2022).
43. Savytskyy, R., Botzem, T., de Fuentes, I. F., Joecker, B., et al.: An Electrically Driven Single-Atom "Flip-Flop" Qubit. Sci. Adv. **9**, eadd9408 (2023).
44. Rei, D., Ferraro, E., De Michielis, M.: Parallel Gate Operations Fidelity in a Linear Array of Flip-Flop Qubits. Adv. Quantum Technol. **5**, 2100133 (2022).
45. Muhonen, J. T., Dehollain, J. P., Laucht, A., Simmons, S., et al.: Coherent Control via Weak Measurements in ^{31}P Single-Atom Electron and Nuclear Spin Qubits. Phys. Rev. B **98**, 155201 (2018).
46. Ban, Y., Kato, K., Iizuka, S., Murakami, S., et al.: Introduction of Deep Level Impurities, S, Se, and Zn, into Si Wafers for High-Temperature Operation of a Si Qubit. Jpn. J. Appl. Phys. **62**, SC1054 (2023).
47. Zhang, S., He, Y., Huang, P.: Acceptor-Based Qubit with Tunable Strain. Phys. Rev. B **107**, 155301 (2023).
48. Partasarathy, S. K., Kallinger, B., Kaiser, F., Berwian, P., et al.: Scalable Quantum Memory Nodes Using Nuclear Spins in Silicon Carbide. Phys. Rev. Appl. 034026 (2023).

49. Pica, G., Lovett, B. W.: Quantum Gates with Donors in Germanium. Phys. Rev. B **94**, 205309 (2016).
50. Madzik, M. T., Asaad, S., Youssry, A., Joecker, B., *et al.*: Precision Tomography of a Three-Qubit Donor Quantum Processor in Silicon. Nature **601**, 348 (2022).
51. Hogg, M. R., Pakkiam, P., Gorman, S. K., Timofeev, A. V., *et al.*: Single-Shot Readout of Multiple Donor Electron Spins with a Gate-Based Sensor. PRX Quantum **4**, 010319 (2023).
52. Pothier, H., Lafarge, P., Urbina, C., Esteve, D., Devoret, M. H.: Single Electron Pump Based on Charging Effects. Europhys. Lett. **17**, 249 (1992).
53. Ruzin, I. M., Chandrasekhar, V., Levin, E. I., Glazman, L. I.: Stochastic Coulomb Blockade in a Double-Dot System. Phys. Rev. B 45, 13469 (1992).
54. Dixon, D. C., Kouwenhoven, L. P., McEuen, P. L., Nagamune, Y., et al.: Influence of Energy Level Alignment on Tunneling Between Coupled Quantum Dots. Phys. Rev. B 53, 12625 (1996).
55. Hu, R.-Z., Ma, R. L., Ni, M., Zhou, Y., et al.: Flopping-Mode Spin Qubit in a Si CMOS Quantum Dot. Appl. Phys. Lett. 122, 134002 (2023).
56. Waugh, F. R., Berry, M. J., Mar, D. J., Westervelt, R. M., et al.: Single-Electron Charging in Double and Triple Quantum Dots with Tunable Coupling. Phys. Rev. Lett. 75, 705 (1995).
57. Veldhorst, M., Ruskov, R., Yang, C. H., Hwang, J. C. C., et al.: Spin-Orbit Coupling and Operation of Multi-Valley Spin Qubits. Phys. Rev. B 92, 201401 (2015).
58. Gamble, J. K., Harvey-Collard, P., Jacobson, N. T., Baczewski, A. D., et al.: Valley Splitting of Single-Electron Si MOS Quantum Dots. Appl. Phys. Lett. 109, 253101 (2016).
59. Neyens, S. F., Foote, R. H., Thorgrimsson, B., Knapp, T. J., et al: The Critical Role of Substrate Disorder in Valley Splitting in Si Quantum Wells. Appl. Phys. Lett. 112, 243107 (2018).
60. Takeda, K., Kamioka, J., Otsuka, T., Yoneda, J., et al.: A Fault Tolerant Adressable Spin Qubit in a Natural Silicon Quantum Dot. Sci. Adv. 2, e1600694 (2016).
61. Leon, R. C. C., Yang, C. H., Hwang, J. C. C., Lemyre, J. C., et al.: Bell-State Tomography in a Silicon Many-Electron Artificial Molecule. Nature Commun. 12, 3228 (2021).
62. Young, S. M., Jacobson, N. T., Petta, J. R.: Optimal Control of a Cavity-Mediated ISWAP Gate Between Silicon Spin Qubits. Phys. Rev. Appl. 18, 064082 (2022).
63. Asai, H., Iizuka, S., Mogami, T., Hattori, J., et al.: Device Structure and Fabrication Process for Silicon Spin Qubit Realizing Process-Variation-Robust SWAP Gate Operation. Jpn. J. Appl. Phys. 62, SC1088 (2023).
64. Langrock, V., Krzywda, J. A., Focke, N., Seidler, I., et al.: Blueprint of a Scalable Shuttle Device for Coherent Mid-Range Qubit Transfer in Disordered Si/SiGe/SiO2. Phys. Rev. X 4, 020305 (2023).
65. Vahapoglu, E., Slack-Smith, J. P., Leon, R. C. C., Lim, W. H., et al.: Single-Electron Spin Resonance in a Nanoelectronic Device Using a Global Field. Sci. Adv. 7, eabg9158 (2021).
66. Wuetz, B. P., Losert, M. P., Koelling, S., Stehouwer, L. E. A., et al.: Atomic Fluctuations Lifting the Energy Degeneracy in Si/SiGe Quantum Dots. Nature Commun. 13, 7730 (2022).
67. Ha, W., Ha, S. D., Choi, M. D., Tang, Y., et al.: A Flexible Design Platform for Si/SiGe Exchange-Only Qubits with Low Disorder. Nano Lett. 22, 1443 (2022).
68. Fang, Y., Philippopoulos, P., Culcer, D., Coish, W. A., Chesi, S.: Recent Advances in Hole Spin Qubits. Mater. Quantum Technol. 3, 012003 (2023).
69. Uddin, W., Khan, B., Dewan, S., Das, S.: Silicon-Based Technology: Progress and Future Prospects. Bull. Mater. Sci. 45, 46 (2022).
70. Spence, C., Paz, B. C., Michal, V., Chanrion, E., et al.: Probing Low-Frequency Charge Noise in Few Electron CMOS Quantum Dots. Phys. Rev. Appl. 19, 044010 (2023).
71. Mickelson, D. L., Carruzzo, H. M., Coppersmith, S. N., Yu, C. C.: Effects of Temperature Fluctuations on Charge Noise in Quantum Dot Qubits. Phys. Rev. B 108, 075303 (2023).

72. Gorman, J., Hasko, D. G., Williams, D. A.: Charge-Qubit Operation of an Isolated Double Quantum Dot. Phys. Rev. Lett. 95, 090502 (2005).

73. Petersson, K. D., Petta, J. R., Lu, H., Gossard, A. C.: Quantum Coherence in a One-Electron Semiconductor Charge Qubit. Phys. Rev. Lett. 105, 246804 (2010).

74. Kim, D., Ward, D. R., Simmons, C. B., Gamble, J. K., et al.: Microwave-Driven Coherent Operation of a Semiconductor Quantum Dot Charge Qubit. Nature Nanotechn. 10, 243 (2015).

75. Petersson, K. D., Smith, C. G., Anderson, D., Atkinson, P., et al.: Microwave-Driven Transitions in Two Coupled Semiconductor Charge Qubits. Phys. Rev. Lett. 103, 016805 (2009).

76. Shi, Z., Simmons, C. B., Ward, D. R., Prance, J. R., et al.: Quantum Coherent Oscillations and Echo Measurements of a Si Charge Qubit. Phys. Rev. B 88, 075416 (2013).

77. Ward, D. R., Kim, D., Savage, D. E., Lagally, M. G., et al.: State-Conditional Coherent Charge Qubit Oscillations in a Si/SiGe Quadruple Quantum Dot. NPJ Quantum Inform. 2, 16032 (2016).

78. Schoenfield, J. S., Freeman, B. M., Jiang, H. W.: Coherent Manipulation of Valley States at Multiple Charge Configurations of a Silicon Quantum Dot Device. Nature Commun. 8, 64 (2017).

79. MacQuarrie, E. R., Neyens, S. F., Dodson, J. P., Corrigan, J., et al.: Progress Toward a Capacitively Mediated CNOT Between Two Charge Qubits in Si/SiGe. NPJ Quantum Inform. 6, 81 (2020).

80. Bashir, I., Blokhina, E., Esmailiyan, A., Leipold, D., et al.: A Single-Electron Injection Device for CMOS Charge Qubits in 22nm FDSOI. IEEE Solid-State Cir. Lett. 3, 206 (2020)

81. Oosterkamp, T. H., Fujisawa, T., van der Viel, W. G., Ishibashi, K., et al.: Microwave Spectroscopy of a Quantum-Dot Molecule. Nature 395, 873 (1998).

82. Lin, T., Xu, Y.-Q., Jiang, S.-L., Wang, N., et al.: Circuit-QED Based Time-Averaged Dispersive Readout of a Semiconductor Charge Qubit. Appl. Phys. Lett. 121, 184004 (2022).

83. Wang, R., Deacon, R. S., Sun, J., Yao, J., et al.: Gate Tunable Hole Charge Qubit Formed in a Ge/Si Nanowire Double Quantum Dot Coupled to Microwave Photons. Nano Lett. 19, 1052 (2019).

84. Rotta, D., De Michielis, M., Ferraro, E., Fanciulli, M., Prati, E.: Maximum Density of Quantum Information in a Scalable CMOS Implementation of the Hybrid Qubit Architecture. Quantum Inform. Process 15, 2253 (2016).

85. Shi, Z., Simmons, C. B., Prance, J. R., Gamble, J. K., et al.: Fast Hybrid Silicon Double-Quantum-Dot Qubit. Phys. Rev. Lett. 108, 140503 (2012).

86. Kim, D., Ward, D. R., Simmons, C. B., Savage, D. E., et al.: High-Fidelity Resonant Gating of a Silicon-Based Quantum Dot Hybrid Qubit. NPJ Quantum Inform 1, 15004 (2015).

87. Maune, B. M., Borselli, M. G., Huang, B., Ladd, T. D., et al.: Coherent Singlet-Triplet Oscillations in a Silicon-Based Double Quantum Dot. Nature 481, 344 (2012).

88. Yang, Y. C., Coppersmith, S. N., Frieson, M.: Achieving High-Fidelity Single-Qubit Gates in a Strongly-Driven Silicon-Quantum-Dot Hybrid Qubit. Phys. Rev. A 95, 062321 (2017)

89. Thorgrimsson, B., Kim, D., Yang, Y.-C., Smith, L. W., et al.: Extending the Coherence of a Quantum Dot Hybrid Qubit," NPJ Quantum Inform. 3, 32 (2017).

90. Ferdous, R., Kawakami, E., Scarlino, P., Nowak, M. P., et al.: Valley Dependent Anisotropic Spin Splitting in Silicon Quantum Dots. NPJ Quantum Inform. 4, 26 (2018).

91. Ferdous, R., Chan, K. W., Veldhorst, M., Hwang, J. C. C., et al.: Interface-Induced Spin-Orbit Interaction in Silicon Quantum Dots and Prospects for Scalability. Phys. Rev. B 97, 241401 (2018).

92. Jang, W., Cho, M.-K., Jang, H., Kim, J., et al.: Single-Shot Readout of a Driven Hybrid Qubit in a GaAs Double Quantum Dot. Nano Lett. 21, 4999 (2021).

93. Ferraro, E., Fanciulli, M., De Michielis, M.: Coherence Time Analysis in Semiconducting Hybrid Qubit under Realistic Experimental Conditions. Adv. Quantum Technol. 1, 1800040 (2018).

94. Ferraro, E., Prati, E.: Is All-Electrical Silicon Quantum Computing Feasible in the Long Term? Phys. Lett. A 384, 126352 (2020.

95. Livadaru, L., Xue, P., Shaterzadeh-Yazdi, Z., DiLabio, G. A., et al.: Dangling-Bond Charge Qubit on a Silicon Surface. New J. Phys. 12, 083018 (2010).

96. Shaterzadeh-Yazdi, Z., Livadaru, L., Taucer, M., Mutus, J., et al.: Characterizing the Rate and Coherence of Single-Electron Tunneling Between Two Dangling Bonds on the Surface of Silicon. Phys. Rev. B 89, 035315 (2014).

97. Bohloul, S., Shi, Q., Wolkow, R. A., Guo, H.: Quantum Transport in Gated Dangling-Bond Aatomic Wires," Nano Lett. 17, 322 (2017).

98. Rashidi, M., Lloyd, E., Huff, T. R., Achal, R., et al.: Resolving and Tuning Carrier Capture Rates at a Single Silicon Atom Gap State. ACS Nano 11, 11732 (2017).

99. Rashidi, M., Vine, W., T. Dienel, Livadaru, L., et al.: Initiating and Monitoring the Evolution of Single Electrons within Atom-Defined Structures. Phys. Rev. Lett. 121, 166801 (2018).

100. Achal, R., Rashidi, M., Croshaw, J., Churchill, D., et al.: Lithogrpahy for Robust and Editable Atomic-Scale Silicon Devices and Memories. Nature Commun. 9, 2778 (2018).

101. Ng, S. S. H., Retallick, J., Chiu, H. N., Lupoiu, R., et al.: SiQAD: A Design and Simulation Tool for Atomic Silicon Quantum Dot Circuits. IEEE Trans. Nanotechnol. 19, 137 (2020).

102. Hall, E.: On a New Action of the Magnet on Electrical Currents. Am. J. Math. 2, 287 (1879).

103. Von Klitzing, K., Dorda, G., Pepper, M.: New Method for High-Accuratacy Determination of the Fine Structure Based on Quantized Hall Resistance. Phys. Rev. Lett. 45, 494 (1980).

104. Aharonov, Y., Bohm, D.: Significance of Electromagnetic Potentials in Quantum Theory. Phys. Rev. 115, 485 (1959).

105. Degen, C. L., Reinhard, F., Cappellaro, P.: Quantum Sensing. Rev. Mod. Phys. 89, 035002 (2017).

106. Zhou, H., Choi, J., Choi, S., Landig, R., et al.: Quantum Metrology with Strongly Interacting Spin Systems. Phys. Rev. X 10, 031003 (2020).

107. Seedhouse, A. E., Tanttu, T., Leon, R. C. C., Zhao, R., et al.: Pauli Blockade in Silicon Quantum Dots with Spin-Orbit Control. Phys. Rev. X Quantum 2, 010303 (2021).

108. Borjans, F., Mi, X., Petta, J. R.: Spin Digitizers for High-Fidelity Readout of a Cavity-Coupled Silicon Triple Quantum Dot. Phys. Rev. Appl. 15, 044052 (2021).

109. Yoon, J., Kim, K., Na, Y., Lee, D.: Characterization and Correction of the Pulse Width Effects on Quantum Sensing Experiments Using Solid-State Spin Qubits. Curr. Appl. Phys. 50, 140 (2023).

110. Hu, Y., Churchill, H. O. H., Reilly, D. J., Xiang, J., et al.: A Ge/Si Heterostructure Nanowire-Based Double Quantum Dot with Integrated Charge Sensor. Nature Nanotechnol. 2, 622 (2007).

111. Stuyck, S. D., Li, R., Kubicek, S., Mohiyaddin, F. A., et al.: An Integrated Silicon MOS Single-Electron Transistor for Spin-Based Quantum Information Processing. IEEE Electron Dev. Lett. 41, 1253 (2020).

112. Zhao, X., Hu, X.: Measurement of Tunnel Coupling in a Silicon Double Quantum Dot Based on Charge Sensing. Phys. Rev. Appl. 17, 064043 (2022).

113. Jin, I. K., Kumar, K., Rendell, M. J., Huang, J. Y., et al.: Combining n-MOS Charge Sensing with p-MOS Hole Double Quantum Dots in a CMOS Platform. Nano Lett. 23, 1261 (2023).

114. Volk, C., Chatterjee, A., Ansaloni, F., Marcus, C. M., Kuemmeth, F.: Fast Charge Sensing of Si/SiGe Quantum Dots via a High-Frequency accumulation Gate. Nano Lett. 19, 5628 (2019).

115. Nakajima, T., Kojima, Y., Uehara, Y., Noiri, A., et al.: Real-Time Feedback Control of Charge Sensing for Quantum Dot Qubits. Phys. Rev. Appl. 15, L031003 (2021).

116. Noiri, A., Takeda, K., Yoneda, J., Nakajima, T., et al.: Radio-Frequency-Detected Fast Charge Sensing in Undoped Silicon Quantum Dots. Nano Lett. 20, 947 (2020).

117. Kamioka, J., Matsuda, R., Mizokuchi, R., Yoneda, J., Kodera, T.: Evaluation of a Physically Defined Silicon Quantum Dot for Design of Matching Circuit for RF Refectometry Charge Sensing. AIP Adv. 13, 035219 (2023).

118. Huang, J. Y., Lim, W. H., Leon, R. C. C., Yang, C. H., et al.: A High-Sensitivity Charge Sensor for Silicon Qubits Above 1 K. Nano Lett. 21, 6328 (2021).

119. Kurashima, K., Kataoka, M., Nakano, T., Fujiwara, K., et al.: Development of Magnetocardiograph without Magnetically Shielded Rooms Using High-Detectivity TMR Sensors. Sensors 23, 646 (2023).

120. Han, C., Xu, M., Tang, J., Liu, Y., Zhou, Z.: Giant Magneto-Impedance Sensor with Working Point Self-Adaptation for Unshielded Human Bio-Magnetic Detection. Vir. Reality and Intell. Hardware 4, 38 (2022).

121. Kanno, A., Nakasato, N., Oogane, M., Fujiwara, K., et al.: Scalp Attached Tangential Magnetoencephalography Using Tunnel-Magnetoresistance Sensors. Sci. Rpts. 12, 6106 (2022).

Optical Qubits

6

Over the past few decades, tremendous interest in optical qubits and systems has arisen, not the least because of the interest in quantum communications systems. Obviously, when one talks about communications over fibers or even through free space, optical methods are implied (most current fiber systems, to the home, as well as for long distances, use quartz fibers and infrared lasers). The connection to nanoelectronics lies in the existence of integrated optical systems; e.g., situations in which the optical system is created on a single chip by normal semiconductor fabrication techniques. However, a complete description of the quantum information world extends beyond just integrated optical chips. In this concept, it is necessary to more fully discuss optical qubits themselves as they may well differ from the forms taken by other qubits. Indeed, often the most basic form lies with the Jaynes-Cummings model already discussed in Chap. 2 [1].

Squeezed states are also important for representing qubits. Measuring these states relies upon the existence of detectors that can sense a single photon without destroying it in the absorption process (so-called non-demolition detectors). These special detectors often rely upon the phase shift that is produced when the photon travels through nonlinear media [2]. Then, entangling the qubits is critical to quantum communications if for no other reason than the desire to encrypt the signals. Needless to say, there are a great variety of integrated optical chips that address a great variety of qubit applications in the optical world. While one goal is to discuss such chips, more important is to understand the key concepts that make optics different from charge- or flux-based systems.

6.1   Squeezed States

The Gaussian wave packet is one of the most common forms of quantum wave functions in optics. This is because it is known to be the minimum uncertainty wave packet [3]. In order to understand what constitutes a "squeezed" state, it is first necessary to understand this minimum uncertainty wave packet. Consider the normalized (the integral of the squared magnitude over space is unity) Gaussian wave packet expressed as

$$\psi(x) = \left(\frac{1}{2\pi\sigma^2}\right)^{1/4} exp\left(-\frac{x^2}{4\sigma^2}\right). \tag{6.1}$$

It is easy to show that the standard deviation (known in quantum mechanics as the uncertainty) for this wave function is $\sqrt{(\Delta x)^2} = \sigma^2$. If this wave function is now Fourier transformed in order to get its equivalent momentum space formulation, one obtains

$$\varphi(k) = \left(\frac{2\sigma^2}{\pi}\right)^{1/4} exp\left(-\sigma^2 k^2\right). \tag{6.2}$$

The uncertainty in momentum (or the standard deviation for this wave function) is given as $\sqrt{(\Delta k)^2} = 1/4\sigma^2$. As a result, the uncertainty in the combined position and wave number is

$$\Delta x \Delta p = \hbar \Delta x \Delta k = \frac{\hbar}{2}. \tag{6.3}$$

The value of this expression is the minimum value allowed by Heisenberg's uncertainly principle [4]. The wave packet's uncertainty in both position and momentum are closely related to one another.

In optics, a propagating *coherent* state can be created by adding the propagation factor to this minimum uncertainty wave packet, as

$$\psi(x) = \left(\frac{1}{2\pi\sigma^2}\right)^{1/4} exp\left(-\frac{x^2}{4\sigma^2} + ip_0 x/\hbar\right). \tag{6.4}$$

Such a propagating minimum uncertainty wave packet is quite often used to represent a single photon and its motion through space, especially in quantum optics. The corresponding momentum space wave packet is now

$$\varphi(k) = \left(\frac{2\sigma^2}{\pi}\right)^{1/4} exp\left(-\sigma^2(k - k_0)^2\right), \quad k_0 = p_0/\hbar. \tag{6.5}$$

It is called a coherent state as the wave packet remains a minimum uncertainty wave packet as it propagates, even though the wave packet broadens in either position or momentum as it propagates; that is, the standard deviation increases with time for a propagating wave packet.

The idea of a *squeezed* state is to reduce the uncertainty of the wave packet [5, 6], by utilizing nonlinear optics. In the first case [5], four-wave mixing was used, and a waveguide is required to have the necessary nonlinearity. In free space, the wave number k and the radian frequency ω is given as

$$k = \frac{\omega}{c}. \tag{6.6}$$

However, in a waveguide, the propagation is slowed down and information flows at a group velocity

$$v_g = \frac{\partial \omega}{\partial k} < c. \tag{6.7}$$

Four-wave mixing usually takes two incoming photons and converts them to two new photons with $\omega_1 < \omega, \omega_2 > \omega, \omega_1 + \omega_2 = 2\omega$. For this to occur one also needs to have the wave number satisfy the same relationships. This means that the dispersion of the waveguide must be nearly linear so that the group velocity is the same for all photons in the process. In the second case [6], parametric down conversion was used to convert photons of frequency ω_1 into a pair of photons at $\omega_2 = \omega_1/2$. We will return to the discussion of waveguides in the section on integrated optics below.

The end result of these processes is to reduce the uncertainty in one of the variables, which results in increasing it in the other variable. In electromagnetics, this is usually measured by the electric and magnetic fields of the optical wave, rather than simply position and momentum. Photons have spin, usually ± 1 corresponding to right-circularly polarized waves and left-circularly polarized waves. If one takes the electric field as the polarization, then this polarization rotates in the right- or left-hand manner as the thumb points in the direction of propagation (the k direction). One convenient convention is to think about the wave amplitude, the electric field, as being X (with 0 phase) and the node value Y (at phase of $\pi/2$) being the out-of-phase component (such as the magnetic field). Then X and Y can be considered to be the conjugate coordinates and the squeezed state is measured with respect to these two variables.

Qubits, and in general non-classical states of light, can be generated by a variety of methods. These can arise from coherent states or squeezed states, both described above. But these, and also cluster states (described more below), tend to be only quasi-quantum states as they tend to be described by Wigner functions with no negative excursions. The Wigner function is determined from the wave function, or more properly from the density matrix of Sect. 2.1 describing the wave function. If this density matrix is written as

$$\rho(r, r') = \psi(r)\psi^\dagger(r'), \tag{6.8}$$

then the Wigner function is found by first transforming to the center of mass coordinates

$$R = \tfrac{1}{2}(r + r'),$$
$$s = r - r'$$
(6.9)

and then performing the Fourier transform on the difference between the position vectors, as

$$W(R, p) = \frac{1}{\hbar^3} \int d^3 s e^{ip \cdot s / \hbar^3} \psi \left(R + \frac{s}{2} \right) \psi^\dagger \left(R - \frac{s}{2} \right).$$
(6.10)

Quantum correlations, such as entanglement, occur in the off-diagonal terms of the density matrix, as described earlier, and these appear as negative excursions of the Wigner function [7]. Hence, the Wigner function is quite often used to evaluate coherence and entanglement in optical quantum systems. Truly non-Gaussian Wigner functions were generated early on from qubits formed from what are called Schrödinger Cat states [8]. More recently, they have been prepared using Rydberg states in an atom cluster [9].

In Fig. 6.1, one approach to achieve squeezed photons is shown. Several optical sources are required in this approach, as two different forms for either a single photon initial wave packet as well as a coherent-superposition wave packet [10]. The initial beam, along with an optical parametric oscillator (OPO) signal are both fed to a beam splitter beam (BS), in which one composite beam is fed to a second beam splitter along with a local oscillator (LO) in order to feed a pair of detectors in which the classical lower frequency gates an electro-optic modulated third beam. The latter is coupled with the second output of the first BS in another BS to measure the squeezed beam. The very large complexity of this optical system is common for the preparation of a squeezed photon beam. This system provides for the homodyne tomography of the initial signals (pink signal generation) and the final output signals. The results of this system are shown in Fig. 6.2. In panel (a), both the phase relationship of the initial and final beams are shown for the single-photon un-squeezed case and for three different amount of squeezing in the signals. Panel (b) illustrates the values of the Wigner functions for these same four situations. The Wigner functions clearly demonstrate the non-classical behavior (with easily observed negative parts) as well as the squeezing of the Wigner function.

Since photon beams of macroscopic intensity can be coherent beams from a laser, one can then argue that the squeezed states are macroscopic quantum states [6]. These states are then candidates for use as qubits. True to its quantum nature, when the photon state is a superposition of several modes, the Wigner distribution develops negative excursions, which are always taken as a sign of entanglement [11]. Such squeezed states have been used to investigate the spatial extent of a single photon [12]. A general treatment of nonlinearity in creating squeezed states has was demonstrated in [13], and a new form of the time evolution operator is used to describe the propagation of the squeezed state [14].

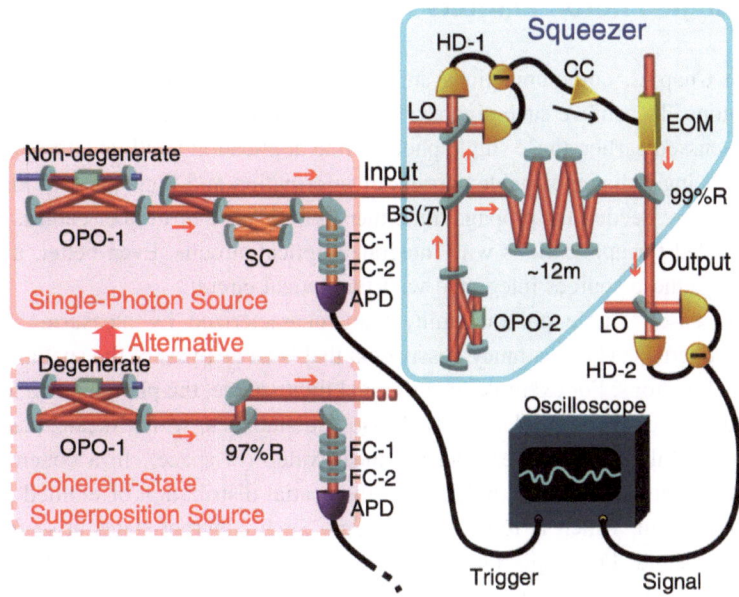

Fig. 6.1 An optical system using single photons (or coherent-state superposition photons) used to create squeezed states. Reprinted with permission form [10], copyright 2014 by the American Physical Society

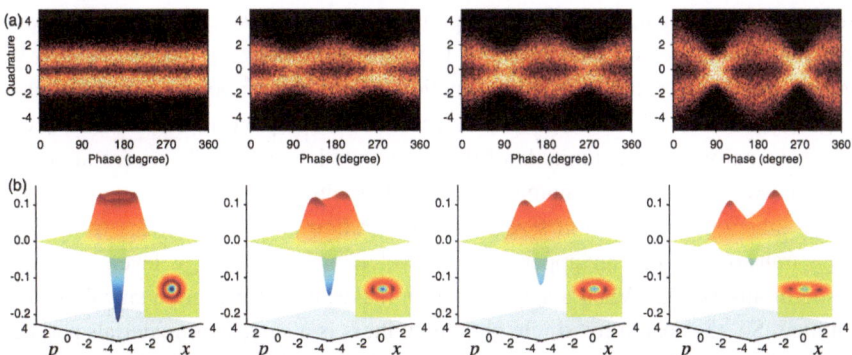

Fig. 6.2 **a** Quantum signals for the single-photon input and the output signals for four situations: non-squeezed on the left, and three other squeezed with squeezing parameter $\gamma = 0.26, 0.37, 0.67$, where $\gamma = -0.5 ln(T)$ and T is the ratio of the two dimensions of the squeezed wave packet. **b** The Wigner functions for the four situations. Reprinted with permission form [10], copyright 2014 by the American Physical Society

6.2   Single-Photon Emitters

Already in Chap. 2, communications and teleportation is discussed in connection with single photons. To achieve such single photons, one needs to have single quantum emitters. As discussed earlier, these single photons find application in all quantum information systems as being indistinguishable carriers of information and logic [15]. In many cases, these sources are needed for propagation either in free space or in fiber cables. But, often they are needed for applications with integrated optical circuits. Even better, it would be useful to have these sources integrated with the optical circuit.

In many cases, single photon emitters are characterized by measuring their anti-bunching properties. This is a fancy description that is an analysis of the spatial separation of individual photons. For example, in a thermal light source, the photons tend to bunched together, and have what is called super-Poissonian statistics, which means that the variance in the distribution is larger than the mean value (in space). In a coherent photon state, such as that discussed in (6.4) above, the spatial distribution of emitted photons is completely random, which is termed Poissonian. Such coherent states arise from lasers far above threshold. For single photon emitters, it is desired to have anti-bunched photons, which have a more regular spatial separation than what comes from a laser. In this case, the statistics are said to be sub-Poissonian, or the variance is smaller than the mean spacing. In general, this variance relates to the fluctuations around the uniform spacing. Basically, a measurement of this may be performed with a 50:50 beam splitter (the optical signal from the emitter is split into a pair beams that propagate normal to each other by the beam splitter). Then, a pair of single photon detectors sense these two beams with the output sent to a correlator that measures the time belay between the two beams. With a single photon detector, the correlation function is zero at zero delay (a single photon can only go to one detector, so the second detector cannot respond if the first one does) and increases to unity as the magnitude of the delay gets larger [16]. This zero at zero delay is the indicator of the single photon nature of the emitter.

As in most optical emission studies, the generation of photons arises from electrons making transitions from an upper energy level to a lower energy level, which is the opposite of absorption. When a single electron makes this transition, a single photon will be emitted. So, the search for single photon emitters is a search for materials and devices which can be controlled in such a manner that only a single electron makes the transition. Thus, the search over the past several decades is actually closely related to finding the set of states that make good qubits. Single atoms and quantum dots are natural sources, as are single molecules and isolated defects within crystals.

Consider the case with quantum dots. Most of the quantum dots discussed in the previous chapters for qubits have energy levels spaced at quite low energies, usually in the radio frequency or microwave frequency, and not in the visible or near-infrared region. The latter is preferred for optical qubits and communications, so different forms of optical dots are preferred. These can be self-organized quantum dots (usually colloidal formed

Fig. 6.3 Schematic representation of the excitation and recombination process for the G center (red transition) in the vicinity of band gap defect states in Si. Reprinted from [19] under the creative commons attribution 4.0 license

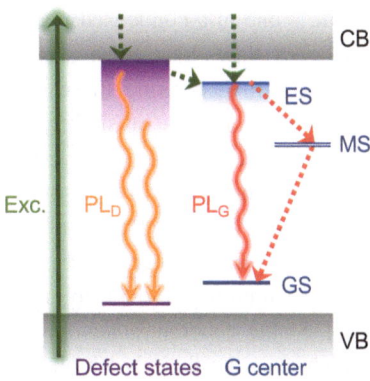

in a liquid suspension) [17], or strain-formed quantum dots formed in epitaxial growth of mismatched heterostructures [18]. In the latter case, a material with a different lattice constant (InAs) is grown on (GaAs), and this results in strain being formed in the epilayer. If less than a monolayer is grown, the strain in this sub-monolayer will lead to it breaking up into a set of localized quantum dot structures of InAs, and these dots have a pyramidal shape. These may then be covered by more GaAs to create an array of quantum dots that make good infrared emitters.

Single photon emitters can be created in silicon by the presence of defects, and thus can be integrated into a Si-based integrated circuit [19]. For example, one such defect is labeled the G-center, which arises from a pair of C atoms sitting on the normal lattice sites with an extra Si atom between them. Another is the W center, that arises from a cluster of three Si atoms sitting at a single lattice site. These tend to be formed by damaging a Si crystal, which is usually produced defect free. In this case, implantation from a focused-ion beam can produce the desired damage. In Fig. 6.3, the transitions involved in emission from the G center are shown [19]. Here, excitation is the green line involving valance band to conduction band transitions. The line PL_D is a normal photoluminescence line to an acceptor level in the carbon rich Si substrate. The ES, MS, and GS states are the excited, metastable, and ground states of the G center, with the red transition being the desired 1.25 μm emission line. The upper ES state must capture an electron from the conduction band to begin the single-photon emission process.

We will see these single quantum emitters in several forms in the following sections dealing with optical qubits and integrated optical circuits.

6.3 Qubits

In the Jaynes-Cummings model of Chap. 2, the qubit was coupled to a resonator. In that case, the qubit was formed from two states in an atom-like structure. But, it was inferred, if not directly stated, that the two-state atom is easily replaced by a real qubit, whether this qubit is one in a condensed matter system or an optical version. In the optical approach, squeezed states have been used for representing qubits or even generating qubits. One approach to generating these states is by four-wave mixing in a microwave cavity, in which two photons of slightly differing frequencies are used to initiate the nonlinear interaction, as discussed above [20].

Other approaches used squeezed states with a nonlinear interaction between the particle-like squeezed photon and a wave-like pair of coherent states [10]. One such approach uses a quantum dot situated within the optical cavity, as a variation on the Jaynes-Cummings model, to create qubit photons that then may be processed elsewhere [21]. Stable coherent states have also been prepared using a Kerr Parametric Oscillator, which uses the third-order nonlinearparametric generation discussed in Sect. 6.1 above [22]. The dynamics of coherent state qubits have been examined as well [23].

Several types of qubits: single-photon qubits, a hybrid qubit formed between a single-photon state and a coherent state, and a single-photon qubit and a state formed from the combination of coherent states, were examined for their efficacy [24]. Single-photon qubits, in which the qubit state is represented by the polarization of the light wave have been created from a wavelength conversion process (such as the four-wave mixing or parametric conversion) and used to analyze long distance communications [25]. A similar long distance study has been made with entangled qubits obtained through Gaussian correlated photonic beams [26]. In this latter study, the two beams were created in a common nonlinear parametric amplifier and then entanglement between photon qubits of each beam was studied.

Entangled photon pairs have been obtained both from emission from a Rb atom [27] as well as from nonlinear interactions in a lithium niobate crystal [28]. They have also been generated using the nonlinear interaction that results from parametric down conversion of signals [29]. In the latter case, the photon pairs are spatially separated either longitudinally or transversely, and generated in a fiber-based system, illustrated in Fig. 6.4 [29]. Here, the input from a weakly-coherent source (WCS) is fed to an optical circulator (C), with the output going to a bi-directional fiber coupler (FC1), which in turn is coupled to a Sagnac loop, a loop that contains a lithium niobate phase modulator that produces a phase ϕ_R in the return signal to the FC1. This produces the two outputs denoted as $|0\rangle$ and $|1\rangle$. The latter is then given an arbitrary phase shift ϕ_x to produce the qubit, such as that of (2.10). These two photons are then converted by a photonic lantern–a construct that allows one to convert a multi-mode system to a single-mode system and vice versa [30]. In the present case, it combines the two input bits into the spatially resolved pair of beams $|LP11a\rangle$ and $|LP11b\rangle$ whose spatial profiles are plotted at the bottom right of the figure. These are then fed into the few-mode fiber (FMF). Photons have also been created via time binning so that they arrive as slightly different times [31–33].

Fig. 6.4 The transversal
spatial state generator. Details
are discussed in the text.
Reprinted from [29] under the
creative commons attribution
4.0 license

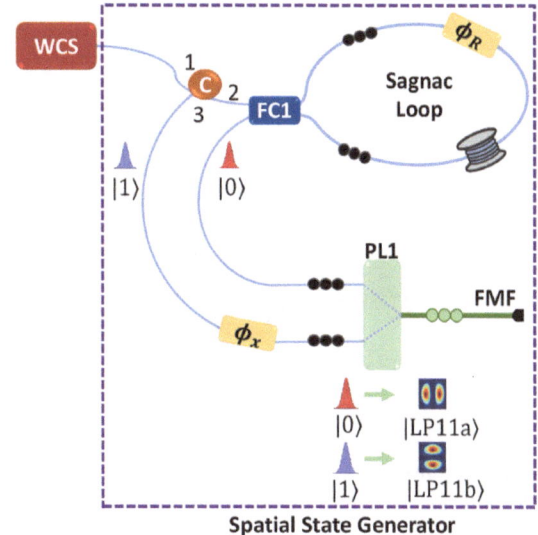

Spatial State Generator

6.4   Optical Gates

Generation of optical gates can be quite expansive in terms of experimental apparatus, as
may be induced from the previous discussion of squeezed states and optical qubits. Even
such a simple gate as the CNOT gate, pictured schematically in Fig. 2.4 of that earlier
chapter, can be quite difficult to achieve optically. An example is shown in Fig. 6.5 [34].
On the left hand side, panel (a), is depicted the creation of the optical qubit from a
Rydberg atom in a cavity (a version of the Jaynes-Cummings model of Chap. 2). In this
case, the atom is an ^{87}Rb atom in the dipole optical cavity, which is excited initially by
the excitation pulse entering from the left. An additional excitation (or even readout) laser
signal (in blue) enters from the right, and the energy levels and transitions for excitation
and readout are shown in panel(a) at the top. The two optical sources are photons of 780
and 479 nm lasers. This produces a set of pulses spaced by about 5 μs. These are fed into
a fiber which leads to an electro-optic modulator (EOM), and then to a beam splitter, so
that a portion of the pulse is sent through a long delay line (lower right fiber coil). This
latter pulse will provide the "control" bit for the CNOT gate. The other signal from the
beam splitter becomes the "target" bit. These two signals are then fed into a set of beam
splitters and mirrors that create a photon-photon interferometer (outlined in green) in
panel(b). The beam splitter within the interferometer is a partial polarization beam splitter
(PPBS), in that it has 33% reflectivity for the horizontal polarization and total reflectivity
for the vertical polarization. When the control bit is in the vertical polarization, the two
photons do not interfere, and the target bit is unchanged. However, when the control bit
is in the horizontal polarization, the target bit goes through a π phase shift and is flipped,
thus producing the controlled NOT operation. The remainder of the circuit (top part) in

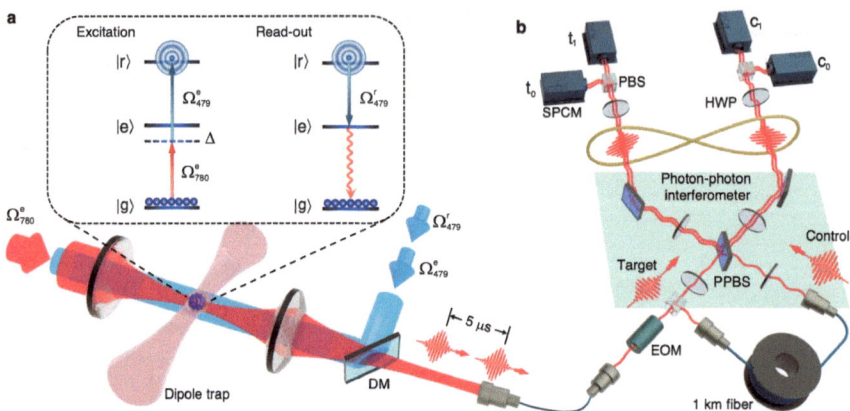

Fig. 6.5 Illustration of the simple CNOT optical gate. The various parts and the discussion the gate operation are described in the text. Reprinted from [34] under the creative commons attribution 4.0 license

panel(b) is the measuring and analysis system to detect whether or not the gate worked. Hence, the heart of the CNOT gate is the interferometer, but the massive remaining parts are for preparation and control of the two qubits, as well as for measuring the results.

Many other gates have been proposed for optical qubits. These include a controlled-Z gate [35–37], a phase gate [38–40], a SWAP gate [41], and even a complete Fourier transform [42]. What all these gates have in common is large optical benches and expansive spatial arrays of optics and lasers, far from the compactness of modern electronics. Nevertheless, an improvement will be seen in the next section in which integrated optical circuits are discussed.

6.5   Integrated Optics

It should not be surprising that a single optical qubit can be extended to multiple qubits and gates of various types, as discussed in the above sections. It also should not be surprising that the optical gates are not as advanced as those for the other technologies that have been discussed in the previous chapters. But, the optical approach has the overwhelming advantage that it is expected to operate efficiently at room temperature, and this advantage should not be overlooked.

Perhaps less obvious is that the optical networks needed to create the qubits and gates described above do not have to be so expansive in spatial terms. Many of these circuits can be integrated in a silicon process that leads to an effective optical processing chip. In some cases, these circuits incorporate optical ring resonators that are coupled to photon sources. These latter sources can be either off the chip or integrated onto it, sometimes as

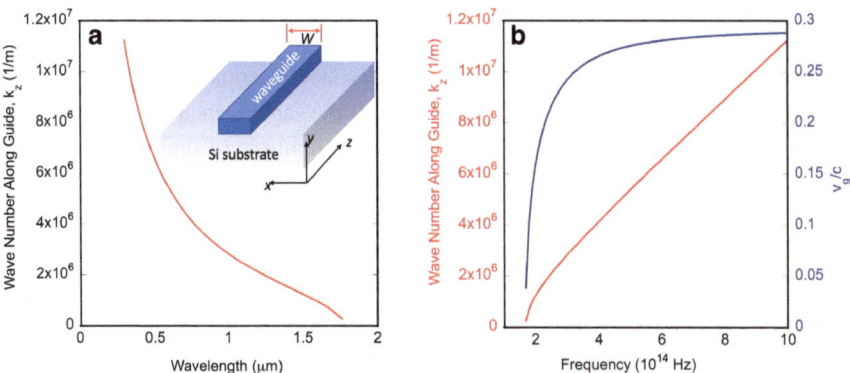

Fig. 6.6   a The propagation wave number k_z along the waveguide as a function of the wavelength. The inset schematic illustrates the guide on its substrate. **b** The propagation wave number k_z along the waveguide and group velocity, as a function of frequency, for an optical wave propagating along the Si waveguide shown in the inset. The assumed width of the guide is $0.5\ \mu m$

single-photon emitters. The ring resonator is a method to enhance the nonlinearity of the silicon waveguides in order to generate four-wave mixing or other parametric processes.

The heart of the integrated optical circuit (IOC) is the waveguide. As discussed above, the waveguide will have a group velocity less than the speed of light, and a cutoff at long wavelength. Generally, the waveguide can be operated in a single-mode manner provided the frequency is kept below the cutoff for higher lying modes of the guide. Such a waveguide in Si is shown schematically in the inset to Fig. 6.6. the waveguide is the darker blue small structure sitting on top of the Si substrate (in light blue). Propagation of the lowest-order mode is determined entirely by the dielectric properties of Si and the width W of the guide.

Generally, the lowest order guided wave is one in which the electric field is vertical, and the wave bounces back and forth between the two sides separated by W. The cutoff at low frequency occurs when the wavelength within the guide is twice the width, or $\lambda = 2W\sqrt{\epsilon_{r1} - \epsilon_{r0}}$, and

$$\omega_0 = \frac{\pi c}{W\sqrt{\epsilon_{r1} - \epsilon_{r0}}}, \tag{6.11}$$

where c is the speed of light in vacuum and ϵ_{r1} and ϵ_{r0} are the relative dielectric values for the waveguide media and free space, respectively. In Si, $\epsilon_{r1} = 11.7$, so that the 0.5 μm waveguide has a cutoff wavelength of 1.7 μm in the guide. The actual free space wavelength for the frequencies of interest and the z-component of wave number, are shown in panel(a) of Fig. 6.6. This wave number along the guide length is given from

$$k_z^2 = k_0^2 \frac{\epsilon_{r1}}{\epsilon_{r0}} - \frac{\pi^2}{W^2(\epsilon_{r1} - \epsilon_{r0})}, \quad k_0 = \frac{\omega}{c}. \tag{6.12}$$

In panel(b), the longitudinal wave number and the group velocity are plotted as a function of the excitation frequency.

As mentioned, the IOC often contains a ring resonator coupled to one or more linear waveguides. Resonance around the ring arises from the same principle as electrons orbiting an atom–the circumference of the ring must be a multiple of the optical wavelength. A sufficiently large ring may have a great many resonances, and one usually requires that a process such as four-wave mixing be induced in a manner that all four frequencies are actual resonances of the ring. Light enters the ring from the waveguide by a *tunneling* process. That is, the optical wave is not completely contained within a dielectric waveguide, as part of it escapes and forms an evanescent wave outside the guide. This wave can couple to another waveguide through the optical process termed frustrated total internal reflection. But, this is basically a semi-classical optical tunneling process of the wave. Nevertheless, the transmitted wave into the ring resonator and the remaining wave in the linear waveguide are entangled, as can be seen from the Wigner function for such a coupling in Fig. 6.7 [43]. Here, the waveguide and the ring resonator are indicated by the black outlines. The width of both waveguides is 0.5 μm, while the outer radius of the ring (of which only one-half is shown) is 5.25 μm. The area depicted in the figure is 13 μm by 13 μm. The coupling distance between the main guide and the ring is 0.25 μm. The laser pulse is taken to be 1.15 fs at 1.5 μm wavelength. In the figure, one can easily see the tunneled pulse in the ring as well as the remainder of the incident pulse in the guide. Between the two lies the oscillatory entanglement–the memory that both pulses came from a single incident pulse. These results have been obtained by a wave function particle-based Monte Carlo approach [44].

While the ring resonator is a common element in an IOC, there are other elements that play a major role. One such element is the Mach–Zehnder Interferometer (MZI). An integrated circuit that highlights the MZI is shown in Fig. 6.8 [45]. In panel(a), the basic circuit is shown. The substrate here is actually GaAs rather than Si, because GaAs is piezo-electric—a fact that will be important later. There are two waveguides, A and B, in both the input and the output. These two guides pass through a pair of multimode interference (MMI) devices that cause interference between the two guides, and form the

Fig. 6.7 Wigner function of an optical pulse that has excited a pulse in the ring resonator. The tunneled pulse and remainder of the incident pulse are shown along with the entanglement of the two. Details are given in the text. Reprinted from [43] with the permission of AIP Publishing

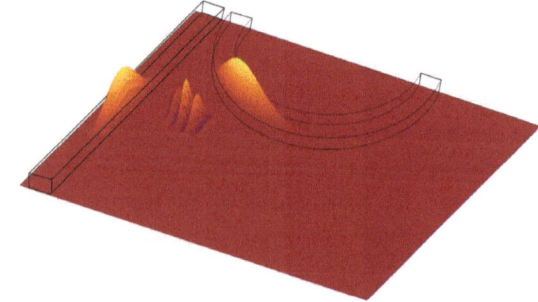

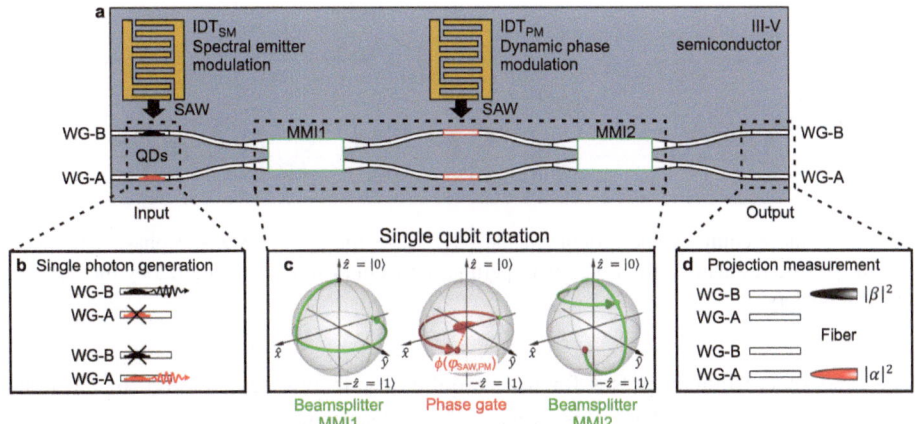

Fig. 6.8 a Schematic representation of the IOC with its inherent MZI. **b** Selective excitation of quantum dots to generate single photons. **c** Rotations on the Bloch sphere, with phase control by the SAW, to produce a phase gate. **d** Collection of the outputs for detection. Reprinted from [45] under the creative commons attribution 4.0 license

MZI between them. The circuit actually is then configured to do more. At the lower left, there are quantum dots that act as single-photon emitters, and are driven by optical signals entering from the left. Then at the top of panel(a) are two surface-acoustic wave (SAW) transducers that use the piezo-electric properties of the substrate to translate microwave signals to surface acoustic waves at 525 MHz. These waves induce atomic vibrations on the substrate surface and these, in turn, are used to tune phase shifters in the waveguides just below them. One is used at the single-photon emitters to modulate the photon's wavelength and the second is used within the MZI to vary the phase of that particular path for qubit operations. This is indicated in panel(c) for operation as a phase gate.

Other basic elements that have been created with waveguides are directional couplers [46], in which a portion of the signal in one waveguide is coupled to a second wave guide, preferentially in a single direction. With the use of phase shifters, ring resonators, MZIs, and directional couplers, a wide variety of optical gates have been integrated onto semiconducting platforms. These include circuits to yield frequency-bin-entangled qubits [47, 48], a CNOT gate [49], a SWAP gate [50], and Fredkin and Toffoli gates [51, 52]. Other IOCs have been made with photonic bandgap materials patterned for waveguides [53, 54] and some insulating glasses [55].

While most of these chips discussed here tend to be limited to one or two gates for an entire chip, there are many suggestions that the IOCs can be improved to contain far more complex circuitry [56]. In Fig. 6.9, such a more complex circuit is shown [48], in which there are 3 MZIs, 3 phase shifters, and 4 small ring resonators. The MZIs allow programming (and/or reconfiguration of the circuit), and this is abetted by the phase shifters. Each of the ring resonators can lead to four-wave mixing to produce a pair of

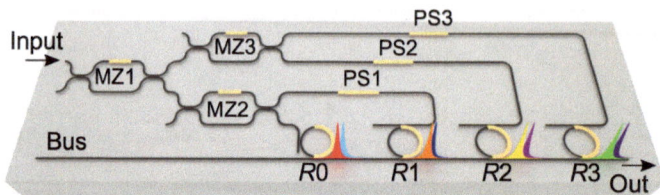

Fig. 6.9 A more complex and programmable IOC. Here, there are 3 MZIs, 3 phase shifters, and four micro-ring resonators. Reprinted with permission from [48], copyright 2023 by the American Physical Society

time (and space) binned pulses for use as qubits (or in this case qudits). In the figure, the first ring is indicated for this process. It is clear that quite complex chips can be created. Nevertheless, the level of optical integration will never approach that of current nanoelectronics, if for no other reason than that photons are significantly larger than the electrons of their counterparts.

6.6   Communications

Quantum communications appears in all forms of quantum information processing. It is required between gates, from sensors to measurement systems, and most of all from one quantum system to another whether they are centimeters apart or hundreds of kilometers apart. As with classical communication systems, multiplexing signals with various techniques enables the channel, be it fiber or over-the-air, to carry far more information. Multiplexing signals, especially for entangled photonic qubits, is never a simple task, but certainly has been studied for both ground and space-based sources and receivers [57]. Such a multiplexing/demultiplexing system is envisioned in Fig. 6.10 [58]. At the top, time-bin entangled photons are produced from a pulsed laser and nonlinear processors. The nonlinearities are in periodically-poled lithium niobite (PPLN) and produce second-harmonic generation (SHG) in one and spontaneous parametric down conversion (SPDC) in the second. These pulses are fed to different channels which provide dense wavelength-division multiplexing (DWDM) according to a standard multiplexing technology standard. Then some of these channels are multiplexed onto three different fiber links and transmitted over an extended distance (30 km in this case). Here, each fiber carries two channels, and these are demultiplexed in the receiver using unbalanced MZIs (UMZI) and sent to their own detectors. The distances can certainly be extended, and communications over several satellite hops have been demonstrated, particularly for key distribution [59].

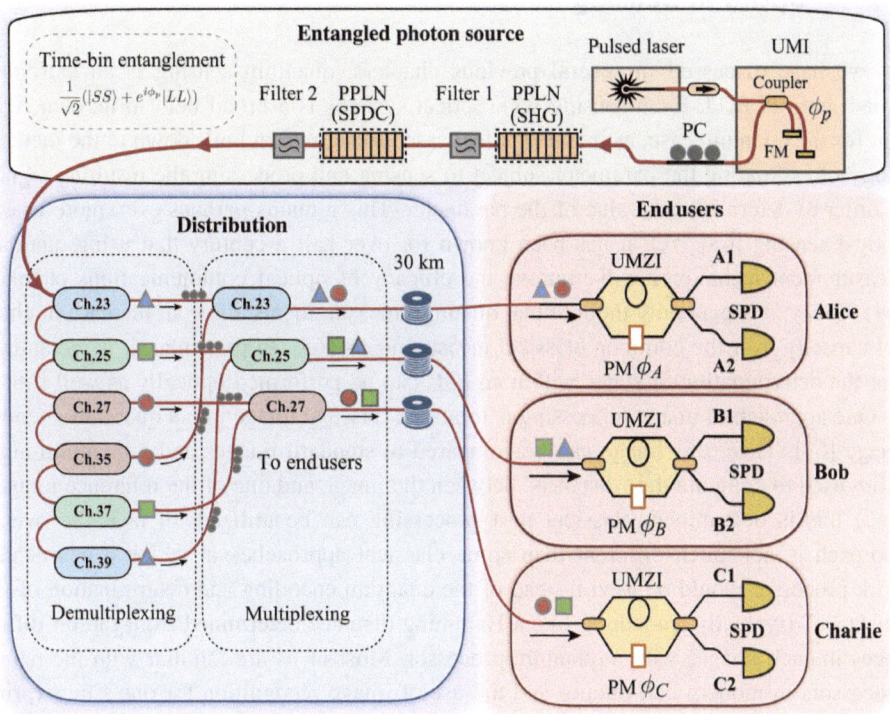

Fig. 6.10 Schematic diagram of the multi-party quantum network in which every member shares the time-bin encoded qubit pairs, in the various channels (Ch.). The network provider consists of the source of the entangled photons and the network nodes based upon the DWDM. Details are discussed in the text. Reprinted from [58] under the creative commons attribution 4.0 license

Such long distance communications have also sent multiplexed signals from photons to quantum dots at the receiver end [60]. In addition, storage of qubits has been studied [61]. In many cases, however, some degree of error correction needs to be built into the system. In one example a five-qubit entangled state was used for the transmission of a two-qubit state [62]. Nevertheless, it is obvious that long distance communications and qubit teleportation are processes best achieved with optical systems. With the quantum key distribution and encoding of qubit discussed in Chap. 2, it is apparent that optical qubits are useful at both the sending and receiving end of the communication channel. Normally, the transmission medium requires the use of optical quantum processors, because optical signals are used today in the distribution of nearly all communications. Nevertheless, whether the signal is microwave, infrared, or visible in nature, it is mostly digital these days.

6.7   Optics in Sensing

As we have discussed in several previous chapters, quantum sensing is an extremely broad, but old field. Even sensing with optical systems is a broad field in its own right. But for the quantum case, as in the classical case, sensing often boils down to the methodology of estimating the parameter subject to sensing and processing the resulting signals in order to determine the value of the parameter. This remains perhaps even more true in optical sensing [63]. Yet, it has been known for over half a century that using quantum measurement techniques will improve the efficacy of optical communications channels [64]. Today, it is generally thought that quantum measurements allow an increase in channel capacity over the common classical measurements [65]. In particular, it is recognized that the determination of states within an IOC can be performed optically as well [66].

One approach to image processing is to map a classical image into a quantum-encoded image [67]. Then, this image can be compared to standard images and a quantum algorithm used to compute the "distance" between the image and one of the reference images. Once this is determined, classical post processing can be utilized. In fact, the overall approach is not much different than some classical approaches in which a neural network processor would be used instead of the quantum encoding and determination of the "distance" (typically something like a Hamming distance determined from the bit differences in each image) with a quantum processor. Most of us are familiar with the neural processors in modern cell phones and the use of image recognition for one's fingerprint, which is just one application of such image processing. In the latter, the neural network processes the differences, looking for a match, but cellular type networks have also been used in the past [68].

Similarly, the use of random-time quantum measurements has been suggested for determining some of the properties of the system that generated the time signal, in particular determining some variables of the Liouvillian of the system [69] (in physics, the Liouvillian is the commutator-generating super-operator form of the Hamiltonian [70], and has somewhat different meanings in mathematics). Here, the similarity is to the use of time-series signal processing to deduce some of the properties of a perhaps chaotic system that is producing the signal [71]. Indeed, such machine learning algorithms are quite often used in image processing as an attempt to extract the image from the noise [72]. Nevertheless, the time-signal processing gives a protocol that is typical for measurement, as shown in Fig. 6.11 [69]. While a linearly polarized beam (say, horizontally polarized) is used to excite the system to be measured, this beam becomes attenuated through passage and also now has both horizontal and vertical components. The beam splitter separates the horizontal and vertical polarizations so that each has its own detector. This allows the experimenter to measure both the attenuation and the phase rotation of the photons.

Perhaps, one of the most direct measurement methods of quantum measurements is the use of optical interferometers. In general, the interferometer is used to measure the change of size or shape of the actual sensing system. An example is the optical sensing

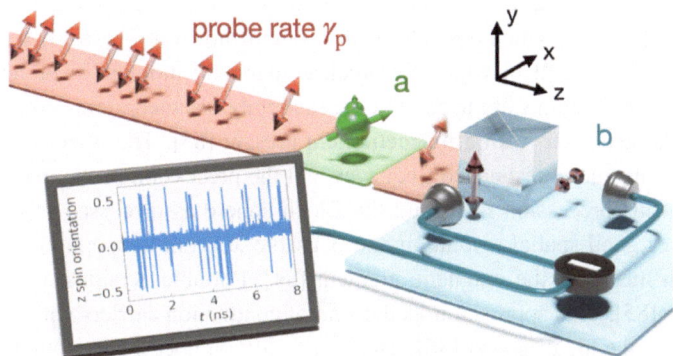

Fig. 6.11 A schematic diagram of random time measurements using optical pulses. The linearly polarized probe photons (top left, in peach) pass through the sample to be measured (green), and then to a beam splitter to reach a pair of detectors (blue region). The detectors then cause either a positive or negative signal (depending upon the detector). Reprinted with permission from [69], copyright 2023 by the American Physical Society

used with the atomic force microscope, in which a fiber interferometer is used to determine movement of the cantilever [73]. In this case, one mirror is actually located on the cantilever, and the fiber connects this to the second mirror and laser source. Another use is to measure temperature through the elongation of a fiber by the heat source, where the change in the interferometer properties is used to deduce the temperature change [74]. These are both methods of using macroscopic optics to detect microscopic changes in the system under measurement [75]. Nanowires appear quite often in optical sensing systems, as single-photon sources can easily be attached to, or created in, the nanowire as can single-photon detectors [76]. In this sense, the nanowire plays the role of an integrated waveguide.

The parity operator was introduced already in Chap. 2 following the discussion of the Jaynes-Cummings model. And, the use of this operator in measurements was mentioned there. It turns out that the parity operator is extremely useful in optical metrology. If the goal of quantum approaches to metrology is to improve the sensitivity of any measurement system, then the optical parity operator is a powerful method to have in the toolbox. In general, the parity operator has appeared as a method of distillation of single photons, and of measuring the phase of such photons. Hence, it can be used in interferometers, but also in general quantum measurements that do not destroy the state [77]. And, these approaches are good for static measurements as well as continuous variable measurements [78]. An extensive review of this approach has been published recently [77].

Magnetometry, the measurement and sensing of magnetic fields was discussed in Chap. 3 with regard to the use of SQUIDs as detectors of the fields. It was already remarked there that some newer optical measurements were replacing the use of SQUIDs. For this purpose, there have been some comparisons of these optically-pumped systems with SQUIDs [79, 80]. One such approach is the spin exchange relaxation free (SERF) magnetometer [81, 82], which basically is the system shown in Fig. 6.11, with the system under test an atomic cell in which the nuclear magnetic moment is of interest. In the case of SERF, alkali atoms are used, and the randomly oriented atoms are excited with a circularly polarized laser, with a magnetic field present [83]. The laser is tuned to a particular nuclear magnetic resonance frequency and the precession of the magnetic moment is measured (this precession is around the Bloch sphere as discussed in Chap. 2). This approach gives good spatial resolution and is usable at room temperature. A variation in this approach measures the magnetic gradient [84]. Portable versions of the system are also available [85]. This approach uses the spin of conduction carriers within semiconductors rather than the alkali atoms [86]. These optically-pumped magnetometers have been used in a variety of normal biological sensing as well as for neural [87] and auditory [88] applications.

References

1. Jaynes, E. T., Cummings, F. W.: Comparison of Quantum and Semiclassical Radiation Theories with Applications to the Beam Maser. Proc. IEEE 51, 89 (1963).
2. Yamamoto, Y., Imoto, N., Machida, S.: Amplitude Squeezing in a Semiconductor Laser Using Quantum Nondemolition Measurement and Negative Feedback. Phys. Rev. A 33, 3243 (1986).
3. Ferry, D. K.: Quantum Mechanics--An Introduction for Device Physicists and Electrical Engineers, 3rd Ed. (CRC Press, Boca Raton, 2021).
4. Heisenberg, W.: Über den Anshaulichen Inhalt der Quantentheoretischen Kinematik und Mechanik. Z. Phys. 43, 172 (1927).
5. Slusher, R. E., Holberg, L. W., Yorke, B., Mertz, J. C., Valley, J. F.: Observation of Squeezed States by Four-Wave Mixing in an Optical Cavity. Phys. Rev. Lett. 55, 2409 (1985).
6. Wu, L.-A., Kimble, H. J., Hall, J. L., Wu, H.: Generation of Squeezed States by Parametric Down Conversion. Phys. Rev. Lett. 57, 2520 (1986).
7. Ferry, D. K., Nedjalkov, M.: The Wigner Function in Science and Technology. IOP Publishing, Bristol, UK (2018).
8. Hacker, B., Welte, S., Daiss, S., Shaukat, A., et al.: Deterministic Creation of Entangled Atom-Light Schrödinger Cat States. Nature Photonics 13, 110 (2019).
9. Magro, V., Vaneecloo, J., Garcia, S., Ourjoumtsev, A.: Deterministic Freely Propagating Photonic Qubits with Negative Wigner Functions. Nature Photonics 17,688 (2023).
10. Miwa, Y., Yoshikawa, J., Iwata, N., Endo, M., et al.: Exploring a New Regime for Processing Optical Qubits: Squeezing and Unsqueezing Single Photons. Phys. Rev. Lett. 113, 013601 (2014).
11. Zayed, E. M. E., Daoud, A. S., Al-Laithy, M. A., Naseem E. N.: The Wigner Distribution for Squeezed Vacuum Superposed State. Chaos Sol. Fractals 24, 967 (2004).
12. Kang, Y., Cho, K., Noh, J., Vitullo, D. L. P., et al.: Remote Preparation of Complex Spatial States of Single Photons and Verification by Two-Photon Coincidence Experiment. Optics Exp. 18, 1217 (2010).
13. Albarelli, F., Ferraro, A., Paternostro, M., Paris, M. G. A.: Nonlinearity as a Resource for Nonclassicality in Anharmonic Systems. Phys. Rev. A 93, 032112 (2016).

14. Ren, G., Du, J.-M., Yu, H.-J., Zhang, W.-H.: Evolution of the Coherent State via a New Time Evolution Operator. Optik 127, 3828 (2016).
15. Diguna, L. J., Tjahjana, L., Darma, Y., Zeng, S., et al.: Light-Matter Interaction of Single Quantum Emitters with Dielectric Nanostructures. Photonics 5, 14 (2018).
16. Zwiller, V., Aichele, T., Benson, O.: Quantum Optics with Single Quantum Dot Devices. New J. Phys. 6, 96 (2004).
17. Achermann, M., Petruska, M. A., Crooker, S. A., Klimov, V.: Picosecond Energy Transfer in Quantum Dot Langmuir-Blodgett Nanoassemblies. J. Phys. Chem. B 107, 13782 (2003).
18. Goldstein, L., Glas, F., Marzin, J. Y., Charasse, M. N., Le Roux, G.: Growth by Molecular Beam Epitaxy and Characterization of InAs/GaAs Strained-Layer Superlattices. Appl. Phys. Lett. 47 1099 (1985).
19. Hollenbach, M., Klingner, N., Jagtap, N. S., Bischoff, L., et al.: Wafer-Scale Nanofabrication of Telecom Single-Photon Emitters in Silicon. Nature Commun. 13, 7813 (2022).
20. Aguayo-Alvarado, A. L., Dominguez-Serna, F., De La Cruz, W., Garay-Palmett, K.: An Integrated Photonic Circuit for Color Qubit Preparation by Third-Order Nonlinear Interactions. Sci. Rpts. 12, 5154 (3022).
21. Sharma, A. K., Sharma, R., Sharma, S. K.: Entanglement Generation for Non-Local Qubits Using Quantum Dot Within the Optical Cavity. Opt. Quantum Electron. 55, 631 (2023).
22. Suzuki, Y., Watabe, S., Kawabata, S., Masuda, S.: Measurement-Based Preparation of Stable Coherent States of a Kerr Parametric Oscillator. Sci. Rpts. 13, 1606 (2023).
23. Yang, M. J., Wu, S.-T.: Exact Dynamics for Optical Coherent State Qubits Subject to Environmental Noise. Phys. Rev. A 89, 022301 (2014).
24. Bose, S., Jeong, H.: Quantum Teleportation of Hybrid-Qubits and Single-Photon Qubits Using Gaussian Resources. Phys. Rev. A 105, 032434 (2022).
25. Luo, X.-W., Zhang, C., Novikova, I., Qian, C., Du, S.: Wavelength Conversion for Single-Photon Polarization Qubits Through Continuous-Variable Quantum Teleportation. Phys. Rev. A 105, 052444 (2022).
26. Agusti, J., Minoguchi, Y., Fink, J. M., Rabi, P.: Long-Distance Distribution of Qubit-Qubit Entanglement Using Gaussian-Correlated Photonic Beams. Phys. Rev. A 105, 062454 (2022).
27. Bae, J., Park, J., Yu, Y. J., Noh, H.-R., Moon, H. S.: Polarization-Entangled Photon Pairs from Warm Atomic Ensemble with Magnetic Background Noise. Adv. Quantum Technol. 6, 2200118 (2023).
28. Thyagarajan, K., Lugani, J., Ghosh, S., Sinha, K., et al.: Generation of Polarization-Entangled Photons Using Type-II Double Periodically Poled Lithium Niobate Waveguides. Phys. Rev. A 80, 052321 (2009).
29. Tseng, Y.-C., Wei, Y.-C., Chen, Y.-C.: Efficient Quantum Memory for Photonic Polarization Qubits Generated by Cavity-Enhanced Spontaneous Parametric Downconversion. Opt. Expr. 30, 19946 (2022).
30. Leon-Saval, L., Argyros, A., Bland-Hawthorn, J.: Photonic Lanterns. Nanophotonics 2, 429 (2013).
31. Bouchard, F., England, D., Bustard, P. J., Heshami, K., Sussman, B.: Quantum Communication with Ultrafast Time-Bin Qubits. PRX Quantum 3, 010332 (2022).
32. Zheng, Y., Sharma, H., Borregaard, J.: Entanglement Distribution with Minimal Memory Requirements Using Time-Bin Photonic Qudits. PRX Quantum 3, 040319 (2022).
33. Lo, H.-P., Ikuta, T., Azuma, K., Honjo, T., et al.: Generation of a Time-Bin Greenberger-Horne-Zeilinger State with an Optical Switch. Quantum Sci. Technol. 8, 035003 (2023).
34. Shi, S., Xu, B., Zhang, K., Ye, G.-S., et al.: High-Fidelity Quantum Logic Gate Based on Near-Optimal Rydberg Single-Photon Source. Nature Commun. 13, 4454 (2022).

35. Hacker, B., Welte, S., Rempe, G., Ritter, S.: A Photon-Photon Quantum Gate Based on a Single Atom in an Optical Resonator. Nature 536, 193 (2016).

36. Babushkin, I., Demircan, A., Kues, M., Morgner, U.: Wave-Shape Tolerant Photonic Quantum Gates. Phys. Rev. Lett. 128, 090502 (2022).

37. Su, Q.-P., Bin, L., Zhang, Y., Yang, C.-P.: Simple Realization of Hybrid Controlled-Controlled-Z Gate with Photonic Control Qubits Encoded via Eigenstates of the Photon-Number Parity Operator. Appl. Phys. Lett. 122, 114001 (2023).

38. Su, Q.-P., Zhang, Y., Bin, L., Yang, C.-P.: Efficient Scheme for Realizing a Multiplex-Controlled Phase Gate with Photonic Qubits in Circuit Quantum Electrodynamics. Fron. Phys. 17, 53505 (2022).

39. Meng, H.: Deterministic Linear-Optical Quantum Control Gates Utilizing Path and Polarization Degrees of Freedom. Phys. Rev. A 105, 032607 (2022).

40. Wang, D. S., Dai, D. D., Narang, P.: Tunable Quantum Logic Gate on Photonic Qubits with a Ladder Emitter. Appl. Phys. Lett. 120, 254001 (2022).

41. Du, F. F., Wu, Y.-M., Fan, G., Ma, Z.-M.: High-Fidelity and Low-Cost Hyperparallel Quantum Gates for Photon Systems via Λ-Type Systems. Ann. Phys. 535, 2200507 (2023).

42. Shi, Y., Waks, E.: Quantum Fourier Transform on Photonic Qubits Using Cavity QED. Phys. Rev. A 106, 013709 (2022).

43. Welland, I., Ferry, D. K.: Wigner Functions in Optoelectronics: Wave-Packet Phase Space Monte Carlo Solver for Waveguide-Ring Coupling. J. Appl. Phys. 128, 153102 (2020).

44. Welland, I., Ferry, D. K.: Wave Packet Phase Space Quantum Monte Carlo Method. J. Comp. Electron. 20, 267 (2021).

45. Bühler, D. D., Weiss, M., Crespo-Poveda, A., Nysten, E. D. S., et al.: On-Chip Generation and Dynamics Piezo-Optomechanical Rotation of Single Photons. Nature Commun. 13, 6998 (2022).

46. Katzman, M., Piasetzky, Y., Rubin, E., Barenboim, B., et al.: Robust Directional Couplers for State Manipulation in Silicon Photonic-Integrated Circuits. J. Lightwave Technol. 40, 7634 (2022).

47. Clementi, M., Sabattoli, F. A., Borghi, M., Gianini, L., et al.: Programmable Frequency-Bin Quantum States in a Nano-Engineered Silicon Device. Nature Commun. 14, 176 (2022).

48. Borghi, M., Tagliavacchi, N., Sabattoli, F. A., El Dirani, H., et al.: Reconfigurable Photonic Chip for the Generation of Frequency-bin-Entangled Qudits. Phys. Rev. Appl. 19, 064026 (2023).

49. Feng, L. T., Zhang, M., Xiong, X., Liu, D., et al.: Transverse Mode-Encoded Quantum Gate on a Silicon Photonic chip. Phys. Rev. Lett. 128, 060501 (2022).

50. Chang, X., Chang, K.-C., Xie, Z., Sarihan, M. C., et al.: A Chip-Scale Polarization-Spatial-Momentum Quantum SWAP Gate in Silicon Nanophotonics. Nature Photonics 17, 656 (2023).

51. Li, Y., Wan, L., Zhang, H., Zhu, H., et al.: Quantum Fredkin and Toffoli Gates on a Versatile Programmable Silicon Photonic Chip. NPJ Quantum Inform. 8, 112 (2022).

52. Li, M., Li, C., Chen, Y., Feng, L.-T., et al.: On-Chip Path Encoded Photonic Quantum Toffoli Gate. Photon. Res. 10, 1533 (2022).

53. Chan, M. L., Aqua, Z., Tiranov, A., Dayan, B., et al.: Quantum State Transfer Between a Frequency-Encoded Photonic Qubit and a Quantum-Dot Spin in a Nanophotonic Cavity. Phys. Rev. A 105, 062445 (2022).

54. De, P., Ranwa, S., Mukhopadhyay, S.: Alternate Scheme for Implementation of 3 Qubit Fredkin Gate with Photonic Bandgap Crystal. Opt. Laser Technol. 167, 109804 (2023).

55. Skyrabin, M. M., Kondratayev, I. V., Dyakonov, I. V., Borzenkova, O. V., et al.: Two-Qubit Quantum Photonic Processor Manufactured by Femtosecond Laser Writing. Appl. Phys. Lett. 122, 121102 (2023).

56. Moody, G., Sorger, V. J., Blumenthal, D. J., Juodawlkis, P. W., et al.: 2022 Roadmap for Integrated Quantum Photonics. J. Phys.: Photon. 4, 012501 (2022).

57. Chen, K. C., Dhara, P., Heuck, M., Lee, Y., et al.: Zero-Added Loss Entangled-Photon Multiplexing for Ground- and Space-Based Quantum Networks. Phys. Rev. Appl. 19, 054029 (2023).

58. Kim, J. H., Chae, J. W., Jeong, Y.-C., Kim, Y.-H.: Quantum Communication with Time-Bin Entanglement over a Wavelength-Multiplexed Fiber Network. APL Photon. 7, 016106 (2022).

59. Koegler, M., Horwath, J., Fechtmann, K., Bihn, L. N., et al.: Laser Transmission of Quantum Bits and Multi-Tera-Bits over Multi-Hop Satellite Orbital Constellations. Proc. SPIE 12413: Free-Space Laser Commun. XXXV, 1241039 (2023).

60. Lago-Rivera, D., Rakonjac, J. V., Grandi, S., de Riedmatten, H.: Long Distance Multiplexed Quantum Teleportation from a Telecom Photon to a Solid-State Qubit. Nature Commun. 14, 1889 (2023).

61. Liu, D. C., Li, P.-Y., Zhu, T.-X., Zheng, L., et al.: On Demand Storage of Photonic Qubits at Telecom Wavelengths. Phys. Rev. Lett. 129, 210501 (2022).

62. Liu, X.-F., Li, D.-F., Zheng, Y.-D., Yang, X.-L., et al.: Experimental Realization of Quantum Controlled Teleportation of Arbitrary Two-Qubit State via a Five-Qubit Entangled State. Chin. Phys. B 31, 050301 (2022).

63. Pirandola, S., Bardhan, B. R., Gehring, T., Weedbrook, C., Lloyd, S.: Advances in Photonic Quantum Sensing. Nature Photon. 12, 724 (2012).

64. Helstrom, C. W.: Quantum Detection and Estimation Theory. J. Stat. Phys. 1, 231 (1969).

65. Burenkov, I. A., Jabir, M. V., Polyakov, S. V.: Practical Quantum-Enhanced Receivers for Classical Communications. AVS Quantum Sci. 3, 025301 (2021).

66. Paesani, S., Ding, Y., Santagati, R., Chakhmakhchyan, L., et al.: Generation and Sampling of Quantum States of Light on a Silicon Chip. Nature Phys. 15, 925 (2019).

67. Das, S., Zhang, J., Martina, S., Suter, D., Caruso, F.: Quantum Pattern Recognition on Real Quantum Processing Units. Quantum Mach. Intell. 5, 16 (2023)/

68. Porod, W., Ferry, D. K.: Pattern Recognition in Highly-Integrated Circuits. Patt. Recogn. 18, 179 (1985).

69. Sifft, M., Hägele, D.: Random-Time Quantum Measurements. Phys. Rev. A 107, 052203 (2023).

70. Knezevic, I., Ferry, D. K.: Partial-Trace Free Time-Convolutionless Equation of Motion for the Reduced Density Matrix. Phys. Rev. E 56, 016131 (2002).

71. Sprout, J. C.: Chaos and Time-Series Analysis. Oxford Univ. Press, Oxford, 2003.

72. Da Cunha, C. R., Aoki, N., Ferry, D. K., Lai, Y.-C.: A Method of Finding the Background Potential of Quantum Devices from Scanning Gate Microscopy Data Using Machine Learning. Mach. Learn.: Sci. Technol. 3, 025013 (2022).

73. Rugar, D., Mamin, H. J., Guethner, P.: Improved Fiber Optic Interferometer for Atomic Force Microscopy. Appl. Phys. Lett. 55, 2588 (1989).

74. Peng, Y., Qin, S., Zhang, S., Zhao, Y.: Optical Fiber Quantum Temperature Sensing Based Upon Single Photon Interferometry. Opt. Lasers Engr. 167, 107611 (2023).

75. Ge, W., Jacobs, K., Zubairy, M. S.: The Power of Microscopic Nonclassical States to Amplify the Precision of Optical Metrology. NPJ Quantum Inform. 9, 5 (2023).

76. Chang, J., Gao, J., Zadeh, I. M., El Shaari, A. W., Zwiller, V.: Nanowire-Based Integrated Photonics for Quantum Information and Quantum Sensing. Nanophoton. 12, 339 (2023).

77. Birrittella, R. J., Alsing, P. M., Gerry, C. C.: The Parity Operator: Applications in Quantum Metrology. AVS Quantum Sci. 3, 014701 (2021).

78. Podoshvedov, M. S., Podoshvedov, S. A.: Family of CV States of Definite Parity and Their Metrological Power. Laser Phys. Lett. 20, 045202 (2023).

79. Zahren, S., Mahmoudzadeh, M., Wallois, F., Bertrouni, M., et al.: Performance Analysis of Optically Pumped 4He Magnetometers vs. Conventional SQUIDs: From Adult to Infant Head Models. Sensors 22, 3093 (2022).
80. Marhl, U., Jodko-Vladzinska, A., Brühl, R., Sander, T., Jazbinsek, V.: Transforming and Comparing Data Between Standard SQUID and OPM-MEG Systems. Plus One 17, e0262669 (2022).
81. Boto, E., Holmes, N., Leggett, J., Roberts, G., et al.: Moving Magnetoencephalography Towards Real-World Applications with a Wearable System. Nature 555, 657 (2018).
82. Sander, T., Jodko-Vladzinska, A., Hartwig, S., Brühl, R., Middelmann, T.: Optically Pumped Magnetometers Enable a New Level of Biomagnetic Measurements. Adv. Opt. Technol. 9, 247 (2020).
83. Bao, B., Wang, R., Li, D.: Quantum-Based Magnetic Field Sensors for Biosensing. Sci. Repts. 13, 4623 (2023).
84. Zhang, Y., Tang, J., Cao, L., Zhao, B., et al.: Fast Measurement of Magnetic Gradient Based on Four-Channel Optically Pumped Atomic Magnetometer. Sensors Accuators A: Phys. 361, 114591 (2023).
85. Oelsner, G., Ijsselsteijn, R., Scholtes, T., Krüger, A., et al.: Integrated Optically Pumped Magnetometer for Measurements within Earth's Magnetic Field. Phys. Rev. Appl. 17, 024034 (2022).
86. Geng, R., Mena, A., Pappas, W. J., McCamey, D. R.: Sub-Micron Spin-Based Magnetic Field Imaging with an Organic Light Emitting Diode. Nature Commun. 14, 1441 (2023).
87. Bu, Y., Prince, J., Mojtahed, H., Kimball, D., et al.: Peripheral Nerve Magnetoneurography with Optically Pumped Magnetometers. Fron. Physiol. 13, 798376 (2022).
88. An, K.-M., Shim, J. H., Kwon, H., Lee, J.-H., et al.: Detection of the 40 Hz Auditory Steady-State Response with Optically Pumped Magnetometers. Sci. Repts. 12, 17993 (2022).

Some Other Qubits

7

In the previous few chapters, the most universally used qubits have been described and some discussion of their use in communications and sensing has been given. But, these are just the most common qubits that one can find in use. There are other methods of creating qubits that are being considered, and a few of these may prove fruitful. In this chapter, we discuss a pair of these that seem to hold some promise. And, there also is some commercial interest in these.

7.1   NV Centers

Unlike silicon, which can be grown essentially defect free, many materials that have wide band gaps (≥ 2 eV) have intrinsic defects. Defect centers have been studied for many years. One of the most useful of these is the nitrogen-vacancy (NV) complex, usually in diamond (one of the forms in which carbon crystallizes) [1, 2]. These NV centers exemplify one kind of defect often referred to as a "color center." These are often deep defects (well away from the conduction or valence bands), with rather well-separated ground and excited states. There are a variety of actual N and V configurations within the crystal structure, each of which produces a different range of defect energy levels [3]. The (perhaps) most useful of these is the case where N sits on a carbon lattice site and one of the adjacent lattice sites has the vacancy (lack of an atom). Since N is a group V atom, there will be five dangling bonds around the vacancy, one each from the three neighboring C atoms and two from the N atom. While the center also has been found in GaAs [4], C_3BN [5], and SiC [6–8], diamond remains the most common carrier. Other approaches have centered on the single vacancy in Si and the divacancy (two adjacent empty lattice

© The Author(s), under exclusive license to Springer Nature Switzerland AG 2025   165
D. K. Ferry, *Quantum Information in the Nanoelectronic World*,
Synthesis Lectures on Engineering, Science, and Technology,
https://doi.org/10.1007/978-3-031-62925-9_7

sites) in SiC [9, 10], as well as a Si-vacancy complex in diamond [11, 12] and SiC [10]. The various energy levels for these centers, and their optical transitions, can be in the infrared or visible, which makes them suitable for use in optical qubit applications. The nature of the qubit structure varies, as one may use the coupling of electronic and nuclear spins [13], or the electronic spin may be coupled to a strain in the crystal [14].

Qubits. The normal NV center, described above, is usually discussed in terms of a six-electron model [15]. In this model for the negatively charged center that results from the five dangling bonds (four are the usual number around a vacancy), there is a ground state of an ^{3}A spin triplet, as shown in Fig. 7.1 [16]. Normally, N has five outer shell electrons, two of which fill the 2s levels, while the other three lie in the three 2p levels and possess an angular momentum. These latter three states are usually denoted as the ^{3}A states in defect studies. Due to some internal splitting, the N center has the two states $m_z = \pm 1$ split away from the $m_z = 0$ level by about 2.87 GHz (12.4 μeV). These three states form the ground state of the complex. The excited states are typically the 3s and 4s levels, although many other levels can be involved. These excited states are often denoted as the ^{3}E states. The degeneracy of the $m_S = \pm 1$ can be lifted with an external magnetic field, so that one of these levels used with the $m_S = 0$ level forms a well-defined two-level system for the qubit levels; e.g., the physical realization of the qubit [17]. The spin states can be polarized by exciting an electron to the upper ^{3}E level with a green laser (a wavelength of 532 nm), and this level can be read out by the resulting emission (a circularly polarized laser will selectively excite from only one of the angular momentum states). As noted above, the small separation of the two qubit states means that the spin qubit can be manipulated with a microwave signal [8]. This qubit transition can also be used to sense the microwaves [18]. Somewhat more modern approaches also exist to read the spin state in a single shot [19].

The problem with using the lower states, and their splitting by ~ 12 μV, arises from the low temperature of operation. To achieve room temperature, the energy levels need

Fig. 7.1 The electronic energy levels for a nitrogen-vacancy complex in diamond. Here, Ω_R is the Rabi frequency of the qubit. Reprinted from [16] under the creative commons attribution 4.0 license

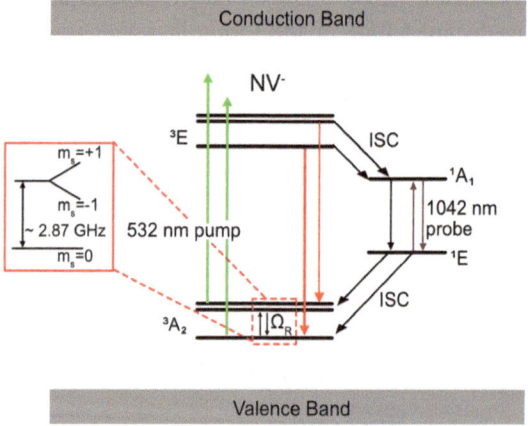

to be separated by something of the order of 100 meV. In cubic SiC, a zero phonon optical transition has been predicted at a wavelength of 1.5 μm (~0.8 eV) [20], which could lead to creating a spin qubit with a relatively long lifetime and with a transition in the near infrared optical regime. Experiments have shown that a line at 1.468 nm is a transition between the excited and ground states of the NV center, which should be useful for quantum communications [21]. The NV center in a nano-diamond crystal was used to demonstrate photon bunching and anti-bunching properties and to be useful for four wave mixing with visible lasers at nitrogen temperatures (~77 K) [22].

If one can achieve operation at higher temperatures, the NV center will be useful as a quantum memory even if the main qubit computation system remains at low temperature. In any case, quantum memories remain a significant problem that some think can be solved with these NV centers. A necessary requirement is to protect the qubit state while also extending the coherence time. Several approaches to achieve this have involved the nuclear spin of the NV center [23], and through decoupling the noise environment from the qubit [24]. Recently, room temperature operation of a two qubit quantum gate using the electron-nuclear entanglement of the nitrogen-vacancy center in diamond has been achieved [25]. In the latter process, the gate functions via microwave control of the geometric phase.

Gates. Naturally, gates have been fabricated using the impurity-vacancy complex in a variety of manners. A silicon-vacancy complex in diamond has been used to create a multi-qubit gate, more for a memory application than for computing itself [26]. An optically excited phase gate has been created by embedding an NV center into the ring of a waveguide-ring resonator configuration [27]. In this case, the optical transition used is from the two angular momentum states of the ^3A level (shown in Fig. 7.1) and a composite state formed from the optically excited states around the ^3E levels. Also, another optical-sensed qubit gate, using the microwave energy levels for the qubit, in which the qubit is manipulated by a sequence of microwave pulses, has been designed for integration with CMOS circuitry [28].

Coupling between the electron spin and the nuclear spin of the N and that of a neighboring C atom has been shown to be another method of performing a quantum algorithm with a single NV center [29]. Other approaches to performing gate operations on a single (or a few) center(s) have also been described [30, 31]. Careful use of different transitions in the energy structure of the NV center also can lead to multiple qubit operations on a single center [32], especially when incorporating the nuclear spin as well [33].

Sensing. As mentioned above, optical readout of the NV center can be quite effective when the center is used as a magnetic field sensor [16]. The NV center, in which the qubit lies in the ^3A level, with the 2.87 GHz separation, has the ability to translate the spin into light field intensity. This gives sensing ability for electromagnetic fields, temperature, spin rotation, and some other areas. For magnetic fields, such as in nuclear magnetic resonance (MRI) or magnetic resonance imaging (MRI), the optically detected magnetic resonance is especially useful [34]. These approaches have been especially useful for biosensing [35],

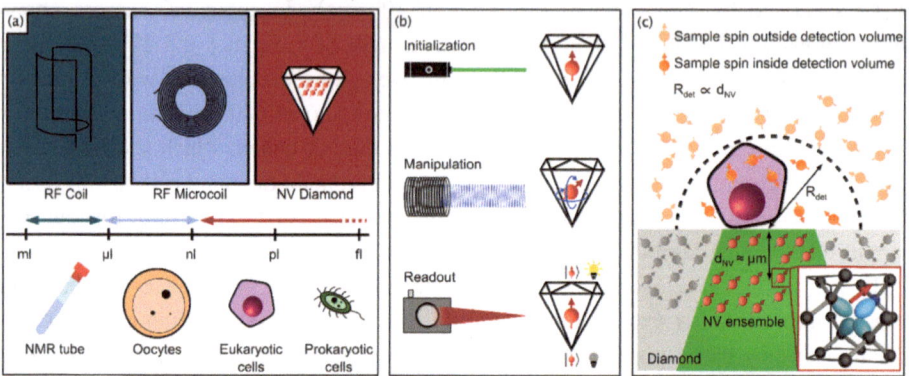

Fig. 7.2 Basic principles of NMR spectroscopy with NV centers in diamond. **a** Comparison of various approaches and the volumes needed for sensing. **b** The various steps of quantum sensing with the NV center. The manipulation involves coupling between the spin of the NV center and that of the cell being measured. **c** The detection radius is indicated by the dashed line. The grey and green areas are the nanodiamond, while the cell to be measured is the purple/red area. Reprinted from [39] under the creative commons attribution 4.0 license

as alternatives to the use of superconducting SQUIDs discussed in Chap. 2, and have found use in magnetoneurography and magnetomyography [36], although there are some frequency limits in sensing of AC magnetic fields [37]. Others, however, have suggested that the frequency is limited only by the clock used with the system (the timing of the controlled qubit transitions) [38]. There is even some evidence for the measurements of NMR in single cells, as shown in Fig. 7.2 [39]. In this latter application, the NV center is located close to the surface of a nanodiamond, and the spin within the biological cell interacts with the spin of the prepared NV center. Readout is optical as described previously. It is thought that this can give a measurement down to the femtoliter volume, basically that of a single small cell [39]. In a similar approach, neurons have been grown on the diamond crystal, and studied directly [40].

Other studies have shown that electric fields can be sensed using the NV center in diamond [41]. It has also been shown that the quantum Fourier transform can be used for sensing with these centers [42].

7.2   Topological Computing

The idea of topological computing rests upon the use of various states that are topologically stable. It is felt that the use of these states will be largely error free, so that dense error correction will not be needed [43]. For example, the edge states (currents flow around the edges of the semiconductor sample due to the quantum Hall effect given by charge accumulation at the edges) of the quantum Hall effect are topologically stable [44],

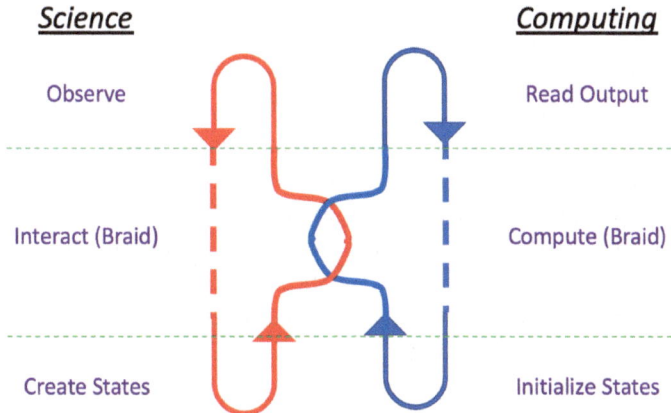

Fig. 7.3 A conceptual view of braiding, as it appears in science and computing. Note how the two colored trajectories are wrapped around each other in the center of the panel. This is the braid operation

as will be discussed further below. In order to achieve computing with these states, the various quantum states must be moved around each other, in a manner termed *braiding*. Quite generally, the concept of braiding is actually connected to the process in which strands of hair are interleaved to form the braids of the hair, and the name is thought to arise from this natural usage. It turns out that there is a mathematical theory for this known as *knot theory* [45]. The application of knot theory in mathematics and physics lies in the analysis of graphs, such as the state transition graphs in computer science mentioned in Chap. 1. As an example, a graph with n vertices leads to a need to evaluate a much larger number of possible arrangements of the system, in which the transition edges are ordered quite differently (with differing interleavings). One must consider all of these in order to compute the partition function of the system itself. The idea of braiding is shown conceptually in Fig. 7.3 for both a scientific calculation and as a computing approach. The red and blue trajectories represent two (presumably quantum) states, and the labels on left and right denote terminology for science and computing. The actual computation gate lies in the center of the figure where the two trajectories are wrapped around one another (braided). This is a simple example, and the various methods of braiding are as numerous as the possible number of knots.

Anyons. The idea of braiding quantum states was recognized quite early as a method for implementing quantum computing, and this is easier with weird quasi-particles[1] that are called *anyons* [46]. It turns out that the separation into bosons or fermions for all particles does not hold in two dimensions, and anyons are generalized quasi-particles that

---

[1] The electron (and the hole) in a semi-conductor is normally a quasi-particle. While it has the same charge as a free electron, its mass is different and depends upon the material in which it is located. Many other quasi-particles arise because of the peculiar science of two dimensions.

can be either fermions or bosons. In our normal three-dimensional world, these particles are *either* fermions *or* bosons (that is, they have either half-integer spin or integer spin, and this doesn't change). But, in two dimensions, the world is much stranger. While one doesn't normally think of two dimensions, in fact in science the world is full of such effectively two-dimensional systems. While a normal semiconductor is three dimensional, the surface of this same semiconductor is two dimensional. The interface, for example between silicon dioxide and silicon in the normal MOSFET, is also two dimensional. It was in this latter interface, of a normal MOSFET at very low temperature, that the quantum Hall effect was discovered [47] (see Sect. 5.7).

In quantized systems such as the quantum Hall effect, the anyon is a quasi-particle that moves in one of the edge states. The term anyon is a generic name that applies to any number of new quasi-particles that can be found in such systems, in which these particles can have fractional statistics. And, fractional statistics are known to occur in the fractional quantum Hall effect [48], so named because the integer ν in (5.8) can actually be a fraction (but a rational fraction).

Because all the quantum Hall states are topologically stabilized, error correction should not be needed in quantum computing [43]. Within this viewpoint, it is expected that these states will appear explicitly as the quasi-particles in the fractional quantum Hall effect [49]. Experiments to explore this were carried out in a unique experiment in which fractional quantum Hall edge states where circulated within an open quantum dot with varying carrying density [44, 50]. Here, the outer edge states were in the 1/3 state while the inner edge states were in the 2/5 state. The nature of the quantum states for these quasi-particles is such that the phase varies with these fractions, so that braiding will cause interference that generates phase differences. Indeed, in the experiment, the quasi-particles of the 1/3 state circulated around those of the 2/5 state, and thus accumulated statistical phase. They found that the phase difference in this process created a shift of $-(2\pi/15)$, clearly demonstrating the correctness of the braiding assumption. (It is probably unnecessary to remind ourselves that it was mentioned in Chap. 1 that Microsoft is pursuing topological quantum computing.) Subsequently, it has been shown that the braiding operation can be done by using an anionic quantum state teleportation, which removes the need to physically transport the anyons around each other [51]. Using lattice models for the anyons has recently been demonstrated [52], and experiments have shown that anyon braiding with 7 superconducting qubits can obtain a high state fidelity in single qubit operations [53].

Majorana Fermions. It turns out that, when one pursues topological computing, it is not enough to pursue just one weird particle, such as the anyon. Some years ago, it was realized that one could effectively stabilize the simplest type of non-Abelian[2] anyon in a computing application by using the topological properties. In particular, one form centers upon a defect that binds *Majorana fermions*. The Majorana fermion is itself a rather

---

[2] A non-Abelian group is one in which any two members of the group do not commute. This non-commutation has been discussed throughout this book as it is a common characteristic of quantum mechanics.

exotic particle. Ettore Majorana hypothesized the existence of a zero-energy particle with half-integer spin as a solution to the Dirac's equations for relativistic quantum mechanics [54]. It was further strange in that it would be its own anti-particle (anti-particles arose in relativity theory, and the positron—or positive electron—is the anti-particle for the electron). (Shortly after this, Majorana apparently disappeared in what has become a scientific mystery.) The mention of the Majorana fermion in connection with a topological insulator material along with proximity superconductivity[3] seems first to have occurred in 2008 [55], although the possibilities of its use in computing is somewhat earlier. When the Majorana fermion is combined with a defect, it becomes a bound state which then is a form of an anyon, with fractional statistics.

Indeed, by combining a "standard" fractional quantum Hall device with a standard superconductor, the resulting structure could support exotic non-Abelian defects with a generalized Majorana bound state [56]. Typical experiments have used proximity induced superconductivity in InSb nanowires (near to a superconducting Al material), as InSb has a large spin–orbit coupling. This allows a large Zeeman splitting between the spins in a magnetic field. Hopefully, when this InSb wire is coupled to a superconductor, signs of the existence of the Majorana fermion will appear [57]. However, braiding operations are not possible in single nanowires since the exchange operation of two Majorana fermions, located at the edge of a nanowire will cause them to annihilate when they meet. Hence, nanowires need to have an extra wire, or leg, section to localize one particle while the other is moved around it [58].

It has been shown subsequently in experiments that a microwave signal can induce Rabi oscillations between adjacent Majorana bound states so that these two states create a single qubit gate [59]. With the introduction of topological insulators,[4] the desired stable states that can be used for qubits are surface states. Since these states occur on the two opposite surfaces of the material (like the opposite two sides of a cube), this can complicate their use. Nevertheless, these topological edge states can be decoupled with "normal" gating operations [60]. Additionally, it has been shown that in the non-Abelian braiding of surface code[5] defects, which can be moved as if they were particles (similar to the Majorana fermions), one can obtain multi-qubit operations [61]. Various novel circuits have been proposed to achieve quantum computing with Majorana fermions [62].

The problem with all of this is that these Majorana particles have not been unambiguously seen in experiments (while they may have been seen in several experiments, there are a variety of other weird quasi-particles that can look very similar). Nevertheless,

---

[3] The proximity effect is when a superconductor can induce superconductivity into a nearby non-superconducting material.

[4] A topological insulator is a material which is an insulator in three dimensions, but produces topologically stable states on its surface (much like quantum Hall states).

[5] Surface codes are various error correcting codes that depend upon the nature of the topological surface in use. One form is the toric code, so named because of the two-dimensional surface being wrapped onto the two torus (a donut shape).

there has been a plethora of theoretical work describing just how they may be manipulated for quantum computing, and other experimental work to examine the detailed physics of the various structures described above for finding these quasi-particles. It is perhaps important that such topological computation schemes have been simulated on both current quantum computers [63–65] and even on classical computers [66].

Other Systems. It is interesting that most of the research on topological computing has focused on the weird and wonderful anyon particles of various sorts. Unfortunately, a great deal of effort has been expended on the search for these particles, and not on progress in actual braiding approaches to quantum computing. However, in the last few years, other approaches to achieving braiding quantum computing have appeared. One approach has used an integrated optical circuit to braid photons [67, 68]. Perhaps of more significance is that the Quantinuum group has used the ion-trap quantum computer to create braiding of the movable ions in a manner that leads to this topological quantum computing [69, 70]. These recent results suggest that perhaps one need not wait for the confirmation and utilization of real Majorana particles. While these are significant steps forward in the braiding approach to quantum computing, it remains unclear whether the error protection these approaches compares with that of the topologically stabilized states.

References

1. Jelezko, F., Wrachtrup, D.: Single Defect Centers in Diamond: A Review. Phys. Stat. Sol. (A) 203, 3207 (2006).
2. Schirhagl, R., Chang, K., Loretz, M., Degen, C. L.: Nitrogen-Vacancy Centers in Diamond: Nanoscale Sensors for Physics and Biology," Ann. Rev. Phys. Chem. 65, 83 (2014).
3. Vimolchalao, S., Liang, W. H., Vila ,F. D., Kas, J. J., et al.: Bethe-Salpeter Equation Calculations of Nitrogen-Vacancy Defects in Damond. J. Phys. Chem. Sol. 123, 87 (2018).
4. Fluegel, B., Beaton, D., Hanna, M., Mascarenhas, A.: Crystallographically Aligned 1.508 eV Nitrogen Pairs in Ultra-Dilute GaAs:N. Jpn. J. Appl. Phys. 57, 090302 (2018).
5. Wang, D., Liu, L., Zhuang, H. L.: Spin Qubit Based on the Nitrogen-Vacancy Center Analog in a Diamond-Like Compound C3BN. J. Appl. Phys. 130, 225702 (2021).
6. Udvarhelyi, P., Gali, A.: Ab Initio Spin-Strain Coupling Parameters of Divancy Qubits in Silicon Carbide. Phys. Rev. Appl. 10, 054010 (2018).
7. Cheng, G. D., Wan, Y. P., Yan, S. Y.: Optical and Spin Coherence Properties of NV Center in Diamond and 3C-SiC. Comp. Mat. Sci. 154, 60 (2018).
8. Zargaleh, S. A., Hameau, S., Eble, B., Margaillan, F., et al. Nitrogen Vacancy Center in Cubic Silicon Carbide: A promising Qubit in the 1.5 μm Spectral Range for Photonic Quantum Networks," Phys. Rev. B 98, 165203 (2018).
9. Son, N. T., Ivanov, I. G.: Charge State Control of the Silicon Vacancy and the Divancey in Silicon Carbide. J. Appl. Phys. 129, 215702 (2021).
10. Fedyanin, D. Y.: Optoelectronics of Color Centers in Diamond and Silicon Carbide: From Single Photon Luminescence to Electrically Controlled Spin Qubits. Adv. Quantum Technol. 4, 2100048 (2021).
11. Berezhnoi, A. D., Zakirov, A. I., Kalachev, A. A.: Quantum Memory Based on SiV-Centers in Nanodiamonds. Laser Phys. Lett. 9, 125206 (2022).

12. Cao, X., Yang, J., Fandrich, T., Zhang, Y., et al.: A Solid-State Source of Single and Entangled Photons at Diamond SV-Center Transitions Operating at 80 K. Nano Lett. 23, 6109 (2023).

13. Gómez, A. V., Rodríguez, F. J., Quiroga, L.: Cross-Entangling Electronic and Nuclear Spins of Distant Nitrogen-Vacancy Centers in Noisy Environments by Means of Quantum Microwave Radiation," Phys. Rev. B 98, 075114 (2018).

14. Udvarhelyi, P., Shkolnikov, V. O., Gali, A., Burkard, G., Pályi, A.: Spin-Strain Interaction in Nitrogen-Vacancy Centers in Diamond," Phys. Rev. B 98, 075201 (2018).

15. Doherty, M. W., Mason, N. B., Delaney, P., Jelezko, F., et al.: The Nitrogen-Vacancy Colour Centre in Diamond. Phys. Repts. 528, 1 (2013).

16. Kim, L., Choi, H., Trusheim, M. E., Wang, H., Englund, D. R.: Nanophotonic Quantum Sensing with Engineered Spin-Optic Coupling. Nanophoton. 12, 441 (2023).

17. Liu, G.-Q., Pan, X.-Y.: Quantum Information Processing with Nitrogen-Vacancy Centers in Diamond. Chin. Phys. B 27, 020394 (2018).

18. Chen, Y., Guo, H., Li, W., Wu, D. W., et al.: Large-Area Tridimenional Uniform Microwave Antenna for Quantum Sensing Based on Nitrogen-Vacancy Centers in Diamond. Appl. Phys. Express 11, 123001 (2018).

19. Hopper, D. A., Shulevitz, H. J., Bassett, L. C.: Spin Readout Techniques of the Nitrogen-Vacancy Center in Diamond," Micromach. 9, 437 (2018).

20. von Bardeleben, H. J., Cantin, J. L., Csóre, A., Gali, A., et al.: NV Centers in 3C,4H, and 6H Silicon Carbide: A Variable Platform for Solid-State Qubits and Nanosensors," Phys. Rev. B 94, 121202 (2016).

21. Cheng, G. D., Wan, Y. P., Yan, S. Y.: Optical and Spin Coherence Properties of NV Center in Diamond and 3C-SiC," Comp. Mat. Sci. 154, 60 (2018).

22. Raza, F., Ahmed, I., Zhang, D., Imran, A., et al.: Bunching and Anti-Bunching in Four Wave Mixing NV Center in Diamond," AIP Adv. 8, 105320 (2018).

23. Pfender, M., Aslam, N., Simon, P., Antonov, D., et al.: Protecting a Diamond Quantum Memory by Charge State Control. Nano Lett. 17, 5931 (2017).

24. Chen, M., Sun, W. K. C., Saha, K., Jaskula, J.-C., Cappellaro, P.: Protecting Solid-State Spins from a Strongly Coupled Environment. New J. Phys. 20, 063011 (2018).

25. Nagata, K., Kuramitan, K., Sekiguchi, Y., Kosaka, H.: Universal Holonomic Quantum Gates over Geometric Spin Qubits with Polarized Microwaves. Nature Comm. 9, 3227 (2018).

26. Stas, P.-J., Huan, Y. Q., Machielse, B., Knall, E. N., et al.: Robust Multi-Qubit Quantum Network Node with Integrated Error Detection. Science 378, 557 (2022).

27. Wang, G. Y., Wei, H.-R.: HIgh-Fidelity Photonic Three-Degree-of-Freedom Hyperparallel Controlled-Phase-Flip Gate. Fron. Phys. 10, 960078 (2022).

28. Omirzakhov, K., Idjadi, M. H., Huang, T.-Y., Breitweiser, S. A., et al.: An Integrated Reconfigurable Spin Control System on 180 nm CMOS for Diamond NV Centers. IEEE Trans. Micro. Theory Techniq. 71, 4052 (2023).

29. Zhang, J., Hegde, S. S., Suter, D.: Efficient Implementation of a Quantum Algorithm in a Single Nitrogen-Vacancy Center in Diamond. Phys. Rev. Lett. 125, 030501 (2020).

30. Yun, M. R., Guo, F.-Q., Yan, L. L., Liang, E., et al.: Parallel-Path Implementation of Nonadiabatic Geometric Quantum Gates in a Decoherence-Free Subspace with Nitrogen-Vacancy Centers. Phys. Rev. A 105, 012611 (2022).

31. Vallabhapurapu, H. H., Hansen, I., Adambukulam, C., Stöhr, R., et al.: High-Fidelity Control of a Nitrogen-Vacancy-Center Spin Qubit at Room Temperature Using the Sinusoidally Modulated, Always Rotating, and Tailored Protocol. Phys. Rev. A 108, 022606 (2023).

32. Bhattacharyya, S., Bhattacharyya, S.: Demonstration of the Holonomically Controlled Non-Abelian Geometric Phase in a Three Qubit System of a Nitrogen-Vacancy Center. Entropy 24, 1593 (2022).

33. Dong, W., Calderon-Vargas, F. A., Economou, S. E.: Precise High-Fidelity Electron-Nuclear Spin Entangling Gates in NV Centers via Hybrid Dynamical Decoupling Sequences. New J. Phys. 22, 073059 (2020).

34. Luo, S., Zhu, Z., Zhang, L., Chen, P., Lin, R.: A Low Complexity Reconstruction Technique for Optical Detection Magnetic Resonance Based Diamond NV Color Center Magnetic Field Measurement. Opt. Quantum Electron. 55, 1004 (2023).

35. Bao, B., Wang, R., Li, D.: Quantum-Based Magnetic Field Sensors for Biosensing. Adv. Quantum Technol. 6, 2200146 (2023).

36. Zhang, C., Zhang, J., Widmann, M., Benke, M., et al.: Optomizing NV Magnetometry for Magnetoneurography and Magnetomyography Applications. Fron. Neurosci. 16, 1034391 (2023).

37. Ghimire, S., Lee, S.-J., Oh, S., Shim, J. H.: Frequency Limits of Sequential Readout for Sensing AC Magnetic Fields Using Nitrogen-Vacancy Centers in Diamond. Sensors 23, 7566 (2023).

38. Boss, J. M., Cujia, K. S., Zopes, J., Degen, C. L.: Quantum Sensing with Arbitrary Frequency Resolution. Science 356, 837 (2017).

39. Neuling, N. R., Allert, R. D., Bucher, D. B.: Prospects of Single-Cell Nuclear Magnetic Resonance Spectroscopy with Quantum Sensors. Curr. Opinions in Biotechnol. 83, 102975 (2023).

40. Losero, E., Jagannath, S., Pezzoli, M., Goblot, V., et al.: Neuronal Growth on High Aspect Ratio Diamond Nanopillar Arrays for Biosensing Applications. Sci. Repts. 13, 5909 (2023).

41. Hollendonner, M., Sharma, S., Parthasarathy, S. K., Dasari, D. B. R., et al.: Quantum Sensing of Electric Field Distributions of Liquid Electrolytes with NV-Centers in Nanodiamonds. New J. Phys. 25, 093008 (2023).

42. Vorobyov, B., Zaiser, S., Abt, N., Meinel, J., et al.: Quantum Fourier Transform for Nanoscale Quantum Sensing. NPJ Quantum Inform. 7, 124 (2021).

43. Rowell, E. C., Wang, Z.: Mathematics of topological quantum computing. Bull. Am. Math. Soc. 55, 183 (2018).

44. Ferry, D. K., Goodnick, S. M., Bird, J. P. Transport in Nanostructures, 2nd. Ed. Cambridge Univ. Press, Cambridge, UK, 2009.

45. Adams, C. C.: The Knot Book: An Introduction to the Mathematical Theory of Knots. W. H. Freeman, New York, 1994.

46. Wilczek, F.: Quantum Dynamics of Fractional-Spin Particles. Phys. Rev. Lett. 49, 957 (1982).

47. Von Klitzing, K., Dorda, G., Pepper, M.: New Method for High-Accuratacy Determination of the Fine Structure Based on Quantized Hall Resistance. Phys. Rev. Lett. 45, 494 (1980).

48. Tsui, D. C., Stormer, H. L., Gossard, A. C.: Two-Dimensional Magnetotransport in the Extreme Quantum Limit. Phys. Rev. Lett. 48, 1559 (1982).

49. Nayak, C., Wilczek, F.: 2n-Quasihole States Realize 2n-1-Dimensional Spinor Braiding Statistics in Paired Quantum Hall States. Nucl. Phys. B 479, 529 (1996).

50. Camino, F. E., Zhou, W., Goldman, V. J.: Realization of a Laughlin Quasiparticle Interferometer: Observation of Fractional Statistics. Phys. Rev. B 72, 075342 (2005).

51. Bonderson, P., Freedman, M., Nayak, C.: Measurement-Only Topological Quantum Computation. Phys. Rev. Lett. 101, 010501 (2008).

52. Nielsen, A. E. B.: Anyon Braiding in Semianalytical Fractional Quantum Hall Lattice Models. Phys. Rev. B 91, 041106 (2015).

53. Song, C., Xu, D., Zhang, P., Wang, J., et al.: Demonstration of Topological Robustness of Anyonic Braiding Statistics with a Superconducting Quantum Circuit. Phys. Rev. Lett. 121, 030502 (2018).

54. Majorana, E.: Teoria Simmetrica dell'Elettrone e del Postrone. Il Nuovo Cim. 14, 171 (1937).

55. Fu, I., Kane, C. L.: Superconducting Proximity Effect and Majorana Fermions at the Surface of a Topological Insulator," Phys. Rev. Lett. 100, 096407 (2008).

56. Clarke, D. J., Alicea, J., Shtengel, K.: Exotic Non-Abelian Anyons from Conventional Fractional Quantum Hall States. Nature Commun. 4, 1348 (2013).
57. Mourik, V., Zuo, K., Frolov, S. M., Plissard, S. R., et al.: Signatures of Majorana fermions in hybrid superconductor semiconductor nanowire devices. Science 336, 1003 (2012).
58. Frolov, S. M., Plissard, S. R., Nadj-Perge, S., Kouwenhoven, L. P., Bakkers, E. P. A. M.: Quantum Computing Based on Semiconductor Nanowires. MRS Bulletin 38, 809–815 (2013).
59. Schmidt, T. L., Nunnenkamp, A., Bruder, C.: Majorana Qubit Rotations in Microwave Cavities. Phys. Rev. Lett. 110, 107006 (2013).
60. Lüpke, F., Just, S., Eschbach, M., Heider, T., et al.: In situ Disentangling Surface State Transport Channel of a Topological Insulator Thin Film by Gating, NPJ Quantum Mater. 3, 46 (2018).
61. Wootton J. R.: Demonstrating Non-Abelian Braiding of Surface Code Defects in a Five Qubit Experiment. Quantum Sci. Technol. 2, 015006 (2017).
62. Lian, B., Sun, X.-Q., Vaezi, A., Qi, X.-L., Zhang, S.-C.: Topological Computation Based on Chiral Majorana Fermions. Proc. Nat. Adad. Sci. 115, 10938 (2018).
63. Huang, H.-L., Narozniak, M., LIang, F., Zhao, Y., et al.: Emulating Quantum Teleportation of a Zero Mode Majorana Qubit. Phys. Rev. Lett. 126, 090502 (2020).
64. Harle, N., Shtanko, O., Movassagh, R.: Observing and Braiding Topological Majorana Modes on Programmable Quantum Simulators. Nature Commun. 14, 2286 (2022).
65. Andersen, T. I., Lensky, Y. D., K, Kechedzhi, K., Drozdov, I. K., et al.: Non-Abelian Braiding of Graph Vertices in a Superconducting Processor. Nature 618, 264 (2023).
66. Zho, D., Pan, N., Chen, T., Sun, H., Zhang, X.: Experimental Simulation of Topological Quantum Computing with Classical Circuits. Adv. Intell. Sys., 2300234 (2023).
67. Zhang, X. L., Yu, F., Chen, Z.-G., Tian, Z.-N., et al.: Non-Abelian Braiding on Photonic Chips. Nature Photon. 16, 390 (2022).
68. Scheel, S., Szameit, A.: A Braid for Light. Nature Photon. 16, 337 (2022).
69. Iqbal, M., Tantivasadakarn, N., Verresen, R., Campbell, S. L., et al.: Creation of Non-Abelian Topological Order and Anyons on a Trapped-Ion Processor. Arxiv.org:2305.03766v1 (2023).
70. Castelvecchi, D.: Physicists Make Long-Sought Topological Quantum States. Nature 617, 445 (2023).

Looking Forward

8

In all of the approaches to quantum information that have been discussed in the previous chapters, it is hard to avoid the fact that one is dealing with nano-electronics (and nano-optics) almost entirely. This should be expected, as it is this industry that has led to the development of the manufacturing technologies that are required to deal with anything that has the word "integrated" in the title. Nevertheless, it can be seen that, of the various quantum technologies discussed, the superconducting technology seems to be the front-runner (or most developed), at least for general qubit generation and sensing today. Now, however, it is time to ask how this technology may evolve in the future. Predicting the future is usually asking for trouble. One cannot know which of these technologies, if any, will evolve to become the standard, even in as few as 5–10 years. This has never been truer than in microelectronics. One may remember that the Si-based microprocessor first appeared in the late 1950s. Devices using GaAs appeared roughly a decade later, with many proposals for it to replace Si for faster computing. But, as an unnamed friend remarked a couple of decades later, "GaAs has been the material of the future for decades, is still the material of the future, and will be the material of the future for several more decades." While GaAs is important for microwave devices, it has never come into the mainstream of microprocessors. Even after GaAs, there were other new materials proclaimed to be the successor to Si. These included (in recent times) carbon nanotubes, graphene, and other two-dimensional materials such as the transition-metal dichalcogenides. Yet, Si trundles along without any real replacement on the horizon, even into the quantum era [1]. Perhaps this is better summarized as saying that wishful thinking is no better at predicting the future than a crystal ball. Having said that, it is not for me to try to predict what the future holds, especially in the quantum information arena.

© The Author(s), under exclusive license to Springer Nature Switzerland AG 2025   177
D. K. Ferry, *Quantum Information in the Nanoelectronic World*,
Synthesis Lectures on Engineering, Science, and Technology,
https://doi.org/10.1007/978-3-031-62925-9_8

What will be done in this chapter is to consider the realm of usage for quantum information, and to describe once more just where the quantum advantage arises, and then to turn to a discussion of a number of possible alternatives and limitations that have appeared in the literature. That is, one should consider this chapter as both a path for expected progress as well as a caution about whether or not some bounds must be overcome, particularly as some negative views are discussed. However, one must be careful in this attempt as it is not at all clear that claimed advantages, applications, and limitations are not just wishful thinking. But, I will leave it to the readers to decide for themselves about the interpretation to be applied, if for no other reason than that (adapted from what someone once said), progress is in the eye of the beholder.

First consider the question as to whether quantum computing is better than classical computing. If one looks at the performance of today's embodiments of quantum computers, the answer has to be "no". But, that is not a fair comparison, since one is comparing a highly developed, well-scaled technology with an immature technology that is full of promise. This is especially true as the current technology may be near the end of its scaling capabilities since a few critical dimensions have been approaching the atomic scale for some time. This is often interpreted as the end of Moore's Law (discussed in Chap. 1). Then, the question has been phrased as "can quantum computation save it" (Moore's law) [2]? The argument that is presented is that noise is always present and quantum noise is worse than classical noise, particularly since quantum processes are probabilistic processes by their very nature. But, this avoids the need for much more error correction in quantum information. This is clearly recognized in the quantum information field and leads to the result that the number of error correcting qubits is usually larger than the number of actual working qubits in any quantum information process. It is for this reason that topological computing schemes hold interest since they promise to obviate the need for most error correction processes.

It has also been argued that there may be practical limits, as opposed to theoretical limits to quantum computation processes [3]. Here, it was argued that there are likely two practical limitations: (1) quantum mechanics does not offer a complete description of complex physical states, especially those involved in quantum processing, and (2) there may be an information-thermodynamic limit that would forbid efficient complex calculations. In the first case, the incompleteness of quantum mechanics was pointed out by Einstein–Podolsky–Rosen [4], and even admitted by Bohr himself [5]. Yet, this has never been a limit to the use of quantum mechanics over the past century. A great many scientific theories/laws are known to be incomplete, but are still used with great advantage. Indeed, the progress in quantum information does not really require deep understanding of the basics or philosophy of quantum mechanics, as the "shut up and compute" world has clearly shown [6]. The second limitation is far closer to wishful thinking than physical reality. This rule would have to be met by classical computing as well, and the complexity of computations carried out on today's supercomputers put the lie to such a limit. Quantum computation has not yet reached the level of classical computation, which suggests that this is no limit for the foreseeable future.

On the other hand, one should carefully consider the claim that quantum supremacy suggests that there are mathematically hard problems for which a quantum computer is much more efficient at finding solutions than a classical Turing machine [7]. In response, it has been pointed out that careful examination of the algorithm considered in the latter paper actually does not establish any quantum advantage over classical methods [8]. Hess and Philipp go a bit further in considering the claim that the set of quantities defining quantum information is more complex than classical information [9]. They assert that this claim is based upon the Bell inequalities [10], but these latter inequalities are not time dependent, and therefore cannot be applied to quantum algorithms in which time measures the progress of the computation. It has also been pointed out that the Bell inequalities are not really quantum as they appear to have been derived by George Boole in the mid-nineteenth century [11], and may be independent of any physical system [12]. Moreover, there are several demonstrated violations of Bell within classical physics [13]. Then, one has to consider the fact that it seems like the creation of a faster quantum algorithm stimulates a new classical algorithm achieving the same results [14, 15]. The sum of these points seem to indicate that there is no reason for believing that quantum computation is more formidable than classical computation, at least at the present time. So, the advantage has to lie in the future.

There have been some approaches to trying to predict the future for quantum computing. Among many others, Gea-Banacloche and Kish raise the issue of power dissipation being a limit [16]. But, as they themselves point out, nanotechnology has gone through several critical times when power dissipation threatened to end Moore's Law. Indeed, even senior executives at a well-known chip maker predicted (perhaps with tongue in cheek) that the chip temperature was would reach that of the sun's surface in future generations. Each such crisis was met with a total redesign of chip architecture with the specific aim to lower the power dissipation. Even today's chips dissipate far less power than those earlier (mid-90s) chips. But, the problem can still arise in the newest chips, as evidenced by the well-publicized heating problems of the iPhone 15 in late 2023. But, Gea-Banacloche and Kish also point out that there are limits to how low the power per bit can go [16], as was already pointed out in Chaps. 1 and 2, and that this leads to a much larger sensitivity to noise. This issue becomes even more important in considering the need for error correction in quantum information systems. It has been said that quantum computing can be scalable to ever larger sizes provided that error correction can be maintained, and to do so depends upon the current principles of quantum mechanics being sustainable and the noise to be neither too strong nor too correlated [17]. But, the assumptions in this theory must also be true, and they must avoid the fact that the very advantages of quantum computing, namely entanglement, also may well lead to strongly correlated noise [18]. Indeed, the latter author points out that, in the quantum world, the $|0\rangle$ and $|1\rangle$ states are not absolutes and that spin is a notoriously weird variable, in that it is all too agreeable to give "up" and "down" values for any chosen axis. This is just the beginning of the problem of uncertainty (or noise, if you prefer), particularly in quantum mechanics. So, the critical scaling property may not be the qubits, but the level/number of auxiliary qubits that are needed for error correction, and does this number grow beyond all bounds? It is not that current practitioners aren't aware of this problem, but perhaps it still may grow uncontrollably in the future without further adaptations.

Another key point to consider is just how many qubits are needed to create a general purpose quantum computer. Suggestions that one needs millions or billions of qubits are probably pushing a little too far. The answer is more likely to be closer to a classical machine. Numbers like 10^{300}, which are bandied about (along with statements that this is more than the sum of all the particles in the universe), are somewhat off-base. Numbers of this size have been used in computers for a very long time. They are handled with floating point numbers, which have a mantissa and an exponent. And, at least in today's laptops, as well as supercomputers, they are handled with registers in the central processing unit that are typically only 64 bits long. In fact, it is not the *size* of the number that is important, it is the *accuracy* with which you want to know that number. If I want to factor a very large integer, do I need thousands of qubits accuracy, or will 10 significant figures be sufficient. Can I then parse the procedure into smaller segments and still maintain the accuracy? At this point, the question really becomes one of determining just how long we can extend the coherence time; e.g., can the coherence time be sufficiently long to allow the computation before errors destroy the result [19]. The technical inquiry then shifts from the science of the qubit to the operation of the quantum ensemble in its entirety, which includes how the architecture affects the qubit itself.

In 2019, the U. S. National Academies of Science carried out a study on quantum computing and its possible future [20]. The results of this were moderately negative, while still pointing out the need to pursue a significant research effort in the area. In particular, they identified the error correction problem as being very significant, and also concluded that a proper quantum computer able to break a 2048-bit encryption popularly in use would require a many decade development effort. Dyakonov went a little further [21], in asking when such a computer would appear. He concluded that this machine would require the control of 2^N quantum amplitudes with $N \sim 10^3 - 10^5$, and that this would occur when "...physicists and engineers learned to control this number of degrees of freedom, which means–*never*." This seems to be too negative an outlook, since it avoids the fact that far more than 10^3 degrees of freedom can be controlled today. A modern parallel computer can use what are called "very long instruction words" which are control words with more than 10^3 bits. Each such word contains separate instructions for each of the parallel processors, the latter of which may all be on a single chip. And, there may be hundreds of these words in the instruction set. Although this architecture is not currently very popular, it does establish the existence of such control, in a way that would, on the surface, satisfy the above constraint.

As was remarked in Chap. 2, the true advantage of quantum information over classical information is entanglement, or quantum parallelism. Without entanglement, one is left with just the analog nature of the qubit, which means it is not limited to just $|0\rangle$ and $|1\rangle$, but can be anywhere in between (which means anywhere on the Bloch sphere). Without entanglement, one needs to consider how quantum computing would appear on a classical analog computer, as has been suggested [22]. This latter work first considers the limitations and questions about quantum computing, which have been discussed in detail in the previous chapters as well as in the above discussion. The author then turns to a so-called Hilbert space analog computing process. In general, classical analog computers are not Turing machines. Instead, they solve differential equations in time (and not in space or other phase space variables). That is, they attack a type of problem that is somewhat more difficult (and with less temporal accuracy) when done on a Turing machine. It is conceptually hard to map a partial-differential equation (with, for example, a number of spatial derivatives) onto an analog computer, which would require a massive interconnected set of parallel analog processors. Here lies the heart of the matter. Quantum mechanics gives rise to partial-differential equations for physical systems. But, the (analog) approach has never been proven not to be meaningful, especially when one asks what the role of entanglement might lead to with the analog approach.

A continuing question that has been asked is whether quantum parallelism, as invoked with entanglement, is a real quantity [23]. The authors then remark that there is an assumption that qubits are being placed into a superposition of various quantities and that this emulates classical parallelism. They then argue that such properties of quantum mechanics are at best speculative. The conclusion drawn from this suggests the advantage of quantum computing is unproven. On the other hand, Schrödinger has argued that entanglement is the most important aspect of quantum theory [24]. Whether entanglement really exists is a critical question, since it has been shown to be necessary in quantum information [25]. Fortunately, there appear to be a number of laboratory experiments confirming the existence of entanglement, and these experiments, and the importance/applications of entanglement, have been the subject of a recent review [26]. Nevertheless, it is well known in science that many so-called proofs are really circular arguments that assume the conclusions, especially in quantum mechanics as has been pointed out by Bell himself [27]. Perhaps this should be a worry, for as stated by Rosinger [12], "...if indeed such may be the case, then that should rather be thoroughly investigated, instead of merely being left to perceptions as part of an attitude which, even if by default, treats it as a taboo."

Much of the above two paragraphs lies more in the realm of philosophy than what we think of as applied physics or, more to the point, applied quantum technology. Beyond just the quantum information world, one might examine our entire existence. The shining of the sun involves nuclear fusion, and the fusion of two atoms depends upon quantum tunneling through the barrier that would normally keep them apart. The actual process first involves entangling the quantum states of the two atoms. This continues over to much of our everyday life, where we deal with applications of nuclear energy, lasers and light absorption in semiconductors, to name just a couple of areas. For the latter, light absorption in a semiconductor, which is important in our solar cells, the entering photon must first produce entanglement between the initial and final states. This means entangling the valence band wave function and the conduction band wave function for the state to which the electron will move in creating the electron and hole (by the absorption of the photon). This entanglement is known as polarization and occurs on a very short time scale (on the order of a few femtoseconds). Yet, it has been explored and measured by fast optical studies using time-resolved four-wave mixing (the same four-wave mixing from Chap. 6) [28]. So, the germane question, in the present context, is not whether entanglement exists in quantum mechanics, but is it being effectively used in today's quantum processors? To answer this question, one needs to look at the current algorithms and applications. There is enough in the previous chapters to suggest that entanglement is occurring and being used in quantum information. But, can one determine just how much this is the case?

It has already been demonstrated that quantum sensing is one of the widest spread applications within quantum information. However, this basically reduces to the measurement problem in quantum mechanics. Measurement generally will break any time symmetry in the system, because it means transferring energy from the quantum system to the classical body that actually displays the result of the measurement [29]. While one hopes that the projection of the quantum state onto the classical body will provide a definitive result, this may not always be the case, as some measurement approaches can lead to chaotic dynamics [30]. While there are many examples given of quantum measurements and sensing in the previous chapters, one still must understand that there are limits to the actual sensing process, even (or, more likely, especially) with quantum methods. The connection between the quantum qubit, and the classical body in which the results of the measurement are displayed, ensure that a connection between quantum and classical states must occur. And, this will entail limits that arise from how the entropy in the two systems is connected, a process that provides both new uncertainty relations [31] and limitations [32]. In fact, the limits on energy required for the measurement do not differ from those discussed in Sect. 2.6. As the world of quantum sensing moves forward, it would be good to keep these limits in mind.

In spite of these precautionary comments, one must still understand that sensing is at the root of both quantum computation and quantum information transfer in addition to the process of measurement itself. This is because, any result from the computation or from the information transfer must also be read by an observer. This observer is a critical part of mainline quantum mechanics [24], but is only a recognition of the classical nature of the final measurement in alternative (and more useful) versions of quantum mechanics [33]. The important point is that sensing cannot be overlooked in any region of quantum information; it is the essential end point and must always be addressed.

What makes a new technology grow is the spread from known applications to new, and perhaps, unknown fields of study. For example, studies with machine learning have pointed out that many processes in machine learning can be transformed into quantum machine learning (QML) with better performance, even on today's machines [34]. This appears to be a result of QML being able to generalize a problem without taking a performance hit. The authors assert an expectation that QML will come to dominate in unsupervised learning as the quantum architecture grows. A similar view is held on the future of quantum sensing technologies [35], with the future promise of advances that overcome technical limitations (such as error correction) in quantum sensing and efforts pushing to more complex applications. In another application, using quantum technologies appears to be opening new approaches to economic models [36]. These are but a few of the areas in which the success of quantum information may well lie in opening new approaches in a variety of fields, rather than merely doing computing faster or better. Just as the suggestion of an information age revolution was created by the microchip [37], it may well be that we are on the verge of the quantum age, in which quantum information technology revolutionizes our way of life once again. But, in many cases, such optimism needs to be tempered in light of the cautions mentioned above, especially as one firm creating quantum AI algorithms has asserted that their algorithms run very well on the classical gpu chips mentioned in Chap. 1 [38].

References

1. Ferry, D. K., Weinbub, J., Nedjalkov, M., Selberherr, S.: A Review of Quantum Transport in Field-Effect Transistors. Semicond. Sci. Technol. **37**, 043001 (2022).
2. Kish, L. B.: Moore's Law is Killed by Classical Physics: Can Quantum Information Save It? Proc. SPIE **5115**, 167 (2003).
3. Paraoanu, G. S.: Quantum Computing: Theoretical versus Practical Possibility. Phys. Perspect. **13**, 359 (2011).
4. Einstein, A., Podolsky, B., Rosen, N.: Can Quantum-Mechanical Description of Physical Reality be Considered Complete? Phys. Rev. **47**, 777 (1935).
5. Bohr, N.: Causality and Complementarity. Phil. Sci. **4**, 289 (1937).
6. Oriols, X., Ferry, D. K.: Why Engineers are Right to Avoid the Quantum Reality Offered by the Orthodox Theory. Proc, IEEE **109**, 955 (2021).
7. Deutsch, D., Jozsa, R.: Rapid Solutions of Problems by Quantum Computation. Proc. Roy. Soc. London A **439**, 553 (1992).
8. Kish, L. B.: "Quantum Supremacy" Revisited: Low-Complexity Deterministic Solutions of the Original Deutsch-Jozsa Problem in Classical Physical Systems. Roy. Soc. Open Sci. **10**, 221327 (2023).
9. Hess, K., Philipp, W.: Quantum Computing and the Theorem of Bell. Microelectron. Engr. **63**, 11 (2002).
10. Bell, J. S.: On the Einstein-Podolsky-Rosen Paradox. Phys. **1**, 195 (1964).
11. Boole, G.: The Laws of Thought. London, Walton and Maberly, 1854.
12. Rosinger, E. E.: George Boole and the Bell Inequalities. arxiv.quant-ph/0406004.

13. Ferry, D. K.: The Copenhagen Conspiracy. Singapore, Pan Stanford, 2019.
14. Tang, E.: A Quantum-Inspired Classical Algorithm for Recommendation Systems. Proc. 51st ACM SIGACT Symp. Theory Comp. STOC 2019, pp. 217–28.
15. Gilyén, A., Song, Z., Tang, E.: An Improved Quantum-Inspired Algorithm for Linear Regression. Quantum **6**, 754 (2022).
16. Gea-Banacloche, J., Kish, L. B.: Future Directions in Electronic Computing and Information Processing. Proc. IEEE **93**, 1858 (2005).
17. Preskill, J.: Sufficient Conditions on Noise Correlations for Scalable Quantum Computing. Quantum Inform. Comp. **13**, 181 (2013).
18. Dyakonov, M. I.: Revisiting the Hopes for Scalable Quantum Computation? JETP Lett. **98**, 514 (2013).
19. Lostaglio, M., Korzekwa, K., Jennings, D., Rudolph, T.: Quantum Coherence, Time-Translation Symmetry, and Thermodynamics. Phys. Rev. X **5**, 021001 (2014).
20. Grumbling E., Horowitz, M., Eds.: Quantum Computing: Progress and Prospects. National Academies of Science, Washington, D. C., 2019.
21. Dyakonov, M. I.: When Will We have a Quantum Computer? Sol. State Electron. **155**, 4 (2019).
22. Kish, L. B.: Quantum Computing with Analog Circuits: Hilbert Space Computing. Proc. SPIE **5055**, 57 (2003).
23. Lanzagorta, M., Uhlmann, J.: Is Quantum Parallelism Real? Proc. SPIE **6976**, 69760W (2008).
24. Schrödinger, E.: Die gegenwartige Situation in der Quantenmechanik. Naturwiss. 23, 803, 827, 844 (1935); Tr. Trimmer, J. D.: The Present Situation in Quantum Mechanics: A Translation of Schrödinger's "Cat Paradox" Paper. Proc. Am. Phil. Soc. **124**, 323 (1980).
25. Linden, N., Popescu, S.: Good Dynamics versus Bad Kinematics: Is Entanglement Needed for Quantum Computation? Phys. Rev. Lett. **87**, 047901 (2001).
26. Horodecki, R., Horodecki, P., Horodecki, M., Horodecki, K.: Quantum Entanglement. Rev. Mod. Phys. **81**, 865 (2009).
27. Bell, J. S.: On the Problem of Hidden Variables in Quantum Mechanics. Rev. Mod. Phys. **38**, 447 (1966).
28. Haug, H., Jauho, A.-P.: Quantum Kinetics in Transport and Optics of Semiconductors. Springer, Berlin, 2004.
29. Bednorz, A., Franke, K., Belzig, W.: Noninvasiveness, Time Symmetry, and Weak Measurements. New J. Phys. **15**, 023043 (2013).
30. Everitt, M. J., Munro, W. J., Spiller, T. P.: Quantum Measurement with Chaotic Apparatus. Phys. Lett. A **28**, 2809 (2010).
31. Korzekwa, K., Lostaglio, M., Jennings, D., Rudolph, T. P.: Quantum and Classical Entropic Uncertainty Relations. Phys. Rev. A **89**, 0422122 (2014).
32. Frenzel, M. F., Jennings, D., Rudolph, T. P.: Reexamination of Pure Qubit Work Extraction. Phys. Rev. E **90**, 052136 (2014).
33. Ferry, D. K.: Quantum Computing and Probability. J. Phys.: Cond. Matt. **21**, 474201 (2009).
34. Tychola, K. A., Kalampokas, T., Papakostas, G. A.: Quantum Machine Learning--An Overview. Electron. **12**, 2379 (2023).
35. Kantsepolsky, B., Aviv, I., Weitzeld, R., Bordo, E.: Exploring Quantum Sensing Potential for Systems Applications. IEEE Access **11**, 31569 (2023).
36. Orrell, D., Houshmand, M.: Quantum Propensity in Economics. Fron. Art. Intell. **4**, 772294 (2022).
37. Evans, C.: The Micromillenium. Washington Square Press, New York, 1982.
38. Wall Street Journal, Sept. 22, 2023.

Appendix A: The Harmonic Oscillator

Almost everybody is familiar with the simple *harmonic oscillator*. This is because the common pendulum is embedded within their consciousness from early in life. This system achieved its name because the motion is harmonic, or has sinusoidal oscillations. It achieves its usage in quantum mechanics because it is one of the few completely solvable problems, at least in the linear limit. Although this common oscillator is a simple system in principle, it achieves almost overwhelming importance due to its usage in describing a plethora of other quantum systems. This is because any complicated interaction potential may be expanded via a Taylor series in which the low-order terms (up to quadratic order) resemble the harmonic oscillator. Thus, the harmonic oscillator becomes the fruit-fly of quantum mechanics. When you understand this simple system you can understand systems that range from a mass on a spring to the world of relativistic quantum electrodynamics.

The sinusoidal motion of this system, with its quadratic potential, is actually quite complicated to solve exactly even in classical mechanics. The differential equation that describes the motion is a well-known boundary value problem, but the solutions are the set of Hermite polynomials, which are not so well known. The standard solution technique is messy and involves a significant number of coordinate transformations. However, quantum mechanics has led to simpler ways to find these solutions, since the quadratic potentials are one of a group of potentials for which the Schrödinger equation can be split into a pair of first-order equations, which provide much deeper insight into the solutions [1, 2]. This approach is actually quicker and easier to understand than the standard solution, and also sets the reader up for the use of operator algebra.

When a simple mass m is suspended by a spring, and the mass is moved away from its equilibrium position, a restoring force tries to move the mass back to its former position. This force may be described as $-Cx$, where x is the displacement of the mass from its equilibrium position and C is known as the spring constant. After moving the mass away from its equilibrium, if the mass is released, it undergoes oscillatory movement, which will only decay if there is some friction in the system (air does the job nicely). This

D. K. Ferry, *Quantum Information in the Nanoelectronic World*,
Synthesis Lectures on Engineering, Science, and Technology,
https://doi.org/10.1007/978-3-031-62925-9

oscillation has a characteristic radian frequency ω, that is usually related to the spring constant as $C = m\omega^2$. The restoring force described above arises from an equivalent potential, which may be written as

$$V = \frac{1}{2}m\omega^2 x^2. \tag{A.1}$$

This potential is be brought into quantum mechanics by introducing it to the Schrödinger equation (time independent form) as

$$-\frac{\hbar^2}{2m}\frac{\partial^2\psi}{\partial x^2} + \frac{1}{2}m\omega^2 x^2\psi = E\psi. \tag{A.2}$$

Here, $\psi(x)$ is the wave function whose solution is sought. The equation is doubly quadratic; that is, it contains one term quadratic in the momentum (quantum mechanically, the momentum $p = -i\hbar\partial/\partial x$ is a differential operator) and one term quadratic in the position. The equation itself may be recognized as the Hermite equation, but the solution will be found using the aforementioned operators.

In order to solve (A.2), a pair of complex adjoint operators are defined in terms of the position and momentum operators as

$$
\begin{aligned}
a &= \sqrt{\frac{m\omega}{2\hbar}}\left(x + i\frac{p}{m\omega}\right) = \sqrt{\frac{m\omega}{2\hbar}}\left(x - \frac{\hbar}{m\omega}\frac{\partial}{\partial x}\right) \\
a^\dagger &= \sqrt{\frac{m\omega}{2\hbar}}\left(x - i\frac{p}{m\omega}\right) = \sqrt{\frac{m\omega}{2\hbar}}\left(x + \frac{\hbar}{m\omega}\frac{\partial}{\partial x}\right)
\end{aligned}. \tag{A.3}
$$

To proceed, it will be necessary to determine the properties of these operators, and these will depend upon the fact that x and p do not commute. In fact,

$$[x, p]\psi = \left(-i\hbar x\frac{\partial\psi}{\partial x} + i\hbar\frac{\partial(x\psi)}{\partial x}\right) = i\hbar\psi. \tag{A.4}$$

(Note that the commutator is often just written as $[x, p] = xp - px = i\hbar$.) This relationship now allows us to write

$$
\begin{aligned}
a^\dagger a &= \frac{m\omega}{2\hbar}\left(x - i\frac{p}{m\omega}\right)\left(x + i\frac{p}{m\omega}\right) \\
&= \frac{1}{\hbar\omega}\left(\frac{p^2}{2m} + \frac{m\omega^2}{2}x^2\right) - \frac{1}{2}
\end{aligned}, \tag{A.5}
$$

and

$$
\begin{aligned}
aa^\dagger &= \frac{m\omega}{2\hbar}\left(x + i\frac{p}{m\omega}\right)\left(x - i\frac{p}{m\omega}\right) \\
&= \frac{1}{\hbar\omega}\left(\frac{p^2}{2m} + \frac{m\omega^2}{2}x^2\right) + \frac{1}{2}
\end{aligned}. \tag{A.6}
$$

Comparing these last two equations, and the original Schrödinger equation (A.2) shows that the energy can already be simplified as

$$E = \hbar\omega\left(a^\dagger a + \frac{1}{2}\right). \tag{A.7}$$

The product of the two operators is not an operator itself, but a simple number which will be shown to be a simple number n, which is the energy level of the particular eigen function of the oscillator. Thus, the harmonic oscillator consists of a set of equally spaced energy levels given as

$$E_n = \hbar\omega\left(n + \frac{1}{2}\right). \tag{A.8}$$

The fact that the energy levels are equally spaced is why the term harmonic has been added to the name. The second piece of useful information is that the operators satisfy their own commutator relationship, which may be expressed from (A.3) as

$$\left[a, a^\dagger\right] = aa^\dagger - a^\dagger a = 1. \tag{A.9}$$

The results (A.8) and (A.9) are the two most important results from the harmonic oscillator. But, to fully understand these, one needs to go further.

Before proceeding, another important property that will be used needs to be explained. The use of the Dirac notation implies an integration as

$$\langle\psi|\psi\rangle = \int\limits_{-\infty}^{\infty} \psi^\dagger\psi\,dx = 1. \tag{A.10}$$

It was stated earlier that the product of the two operators is a simple number, so that it can be expressed as $a^\dagger a = \lambda_n$ (it will be shown below that λ_n is just n). The integration of (A.10) can be used to determine the expectation of this constant as

$$\begin{aligned}\langle\psi|a^\dagger a|\psi\rangle &= \int\limits_{-\infty}^{\infty} \psi^\dagger a^\dagger a\psi\,dx. \\ &= \int\limits_{-\infty}^{\infty} (a\psi)^\dagger a\psi\,dx \geq 0\end{aligned} \tag{A.11}$$

Between the first and second line, an integration by parts has been done–moving the adjoint operator from after the adjoint of the wave to before it requires taking the adjoint of the adjoint. This may be written in Dirac form as

$$\langle a\psi|a\psi\rangle \geq 0. \tag{A.12}$$

So, the result is that the constant is at least positive semi-definite.

At this point, it is necessary to detail the nature of these operators and the properties of the various eigen functions. To begin, consider the product $a^\dagger\psi_n$. To examine this, this product will replace the wave function in the right hand part of (A.11). Then, some processing leads to

$$a^\dagger a\left(a^\dagger \psi_n\right) = a^\dagger \left(aa^\dagger \psi_n\right)$$
$$= a^\dagger \left(a^\dagger a + 1\right)\psi_n = a^\dagger (\lambda_n + 1)\psi_n .$$
$$= (\lambda_n + 1)a^\dagger \psi_n$$

(A.13)

This result says simply that using the operator $a^\dagger$ *increased* the energy in the state by one unit. Now, consider the state $a\psi_n$. Following the same procedure yields

$$a^\dagger a(a\psi_n) = \left(aa^\dagger - 1\right)a\psi_n = (\lambda_n - 1)a\psi_n.$$

(A.14)

This result says simply that using the operator a *decreased* the energy in the state by one unit. There is always a lowest-energy state, which is called the *ground state*. If one tries to lower the energy of this ground state by using the operator a, the result has to be zero, or

$$a\psi_0 = 0.$$

(A.15)

Hence, the results of (A.13), (A.14), and this last equation is that $\lambda_n = n$.

Finally, (A.15) can be used to actually find the wave functions (without having to solve the original Hermite equation). The first operator of (A.3) is used in this last equation to write

$$a\psi_0 = \sqrt{\frac{m\omega}{2\hbar}}\left(x + i\frac{p}{m\omega}\right)\psi_0 = 0.$$

(A.16)

Since p is a differential operator, this first-order equation can be solved to give

$$\psi_0(x) = C_0 exp\left(-\frac{m\omega}{2\hbar}x^2\right).$$

(A.17)

By the properties of Gaussian functions, it is easy to realize that the constant in front must be

$$C_0 = \left(\frac{m\omega}{\pi\hbar}\right)^{1/4}.$$

(A.18)

From the ground state, the wave function for any arbitrary energy level can be found by using the operators as

$$\psi_n \sim \left(a^\dagger\right)^n \psi_0 \sim \left(\frac{m\omega}{2\hbar}\right)^{n/2}\left(x - i\frac{p}{m\omega}\right)^n \psi_0.$$

(A.19)

An equality is not used here, as it is not clear that this wave function is properly normalized for $n > 0$. This may be checked by using the left-hand sides of (A.11) and (A.12) to give the operations

$$\langle a\psi_n | a\psi_n \rangle = \langle \psi_n | a^\dagger a\psi_n \rangle = nC_n^2$$
$$= \langle \psi_{n-1} | \psi_{k-1} \rangle = C_{n-1}^2 \tag{A.20}$$

Hence, the constants for the various levels differ by a square root factor, and (A.18) can be re-expressed as

$$\psi_n = \frac{1}{\sqrt{n+1}} \left(a^\dagger \right)^n \psi_0 = \frac{1}{\sqrt{n+1}} \left(\frac{m\omega}{2\hbar} \right)^{n/2} \left(x - i\frac{p}{m\omega} \right)^n \psi_0. \tag{A.21}$$

Finally, it is useful to have the actual operators x and p in terms of the creation and annihilation operators, and this is given by

$$x = \sqrt{\frac{2\hbar}{m\omega}} \left(a^\dagger + a \right)$$
$$p = i\sqrt{2\hbar m\omega}(a - a) \tag{A.22}$$

References

1. Cooper, F., Khare, A., Sukhatme, U.: Supersymmetry and Quantum Mechanics. Phys. Repts. **251**, 267 (1995).
2. Cooper, F., Khare, A., Sukhatme, U.: Supersymmetry in Quantum Mechanics. (World Scientific, Singapore, 2001).

Appendix B: Simple QED

In the classical world electrodynamics deals with the coupling of photons, whether they are from visible light or microwaves, with electrons. When quantum mechanics is added, it is called quantum electrodynamics, or simply QED in acronym form. The Jaynes-Cummings model is one approach to studying this coupling between photons and electronics, in this case the electrons that exist in the two-level atomic-like dot. It is not the purpose here to review this entire field. Rather, the discussion here is going to be devoted to why the electromagnetic field is quantized and why it is usually represented as a harmonic oscillator set of energy levels.

In Chap. 2, the Jaynes-Cummings model employed an optical cavity, often referred to in Chap. 3 as a resonator. This cavity is shown in Fig. B.1 and will be used in the discussion here. The important point here is that the cavity has a finite length, or width, indicated by W. For this cavity to be a resonator, it is necessary that this length be an exact multiple of one-half the wavelength. The optical wavelength, denoted by λ is distance over which the oscillating signal completes exactly one period of this oscillation. Thus, we require that the length of the cavity be

$$W = \frac{s\lambda}{2},$$ (B.1)

where s is an integer. This simple requirement actually quantizes the electromagnetic wave as a result of the confinement of the wave to the resonator. In essence, this means that the precise wavelength may be give as $\lambda_s = 2W/s$. The corresponding wave number is related to the wavelength (in electromagnetic theory) as

$$k_s = \frac{2\pi}{\lambda_s}.$$ (B.2)

The electromagnetic wave equation which describes the propagation of an electromagnetic wave may be written as [1], in one dimension,

$$\frac{\partial^2 V}{\partial x^2} - \frac{1}{c^2}\frac{\partial^2 V}{\partial t^2} = 0.$$ (B.3)

D. K. Ferry, *Quantum Information in the Nanoelectronic World*, Synthesis Lectures on Engineering, Science, and Technology, https://doi.org/10.1007/978-3-031-62925-9

Fig. B.1 Schematic of a cavity resonator in which the finite length provides spatial quantization of the electromagnetic wave

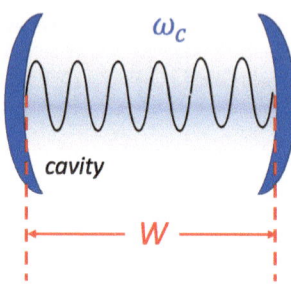

Here, c is the speed of light in the material in which the cavity resides. Typically, a plane wave propagating in the cavity of Fig. B.1 will have the form

$$V \sim exp[i(kx - \omega t)]. \tag{B.4}$$

where the x-axis has been taken as the direction along which the cavity is formed. If this form is used in (B.3), along with the quantization of (B.2), one arrives at

$$\omega_s^2 = c^2 k_s^2. \tag{B.5}$$

Thus, the frequency of the cavity is quantized as well as the wavelength, and this quantization of ω arises from the same spatial effects as (B.1). As a result, the electromagnetic wave in the cavity resonator is now a quantum object, although this spatial quantization is really just a result of classical confinement. The real impact of the quantization will be dealt with in the following.

In many methods of usage, a cavity resonator is depicted as an LC resonator in the circuit schematic, where L is an inductor and C is a capacitor. This was the case in Fig. 3.10 (the red part of the circuit) which is a circuit representation of a transmon qubit. Quite generally, the total energy in the LC circuit may be expressed in Hamiltonian form as [2]

$$H = \frac{Q^2}{2C} + \frac{\Phi^2}{2L}. \tag{B.6}$$

In this equation, Q is the charge residing on the capacitor and Φ is the flux flowing through the inductor. The use of these two variables is governed by the fact that, in relativistic electromagnetics, the charge and the flux are the only two variables that are invariant under any coordinate or time transformation. That is, charge and flux are relativistically invariant. So these are the natural variables to be quantized. This follows from requiring that these two variables do not commute and must satisfy a corresponding quantum commutator as

$$[Q, \Phi] = i\hbar. \tag{B.7}$$

To proceed further in connecting this with harmonic oscillator behavior, one uses the fact that the resonant frequency of the cavity is related to the inductance and capacitance as

$$\omega_s = \frac{1}{\sqrt{LC}}.$$ (B.8)

Thus, it seems that the capacitor and inductance also must have quantized values, although there is some freedom in assigning these values. Nevertheless, if this frequency is introduced into (B.6), a harmonic oscillator type equation is recovered

$$E = H = \frac{\Phi^2}{2L} + \frac{1}{2}L\omega_s^2 Q^2.$$ (B.9)

The equivalent form for the actual harmonic oscillator is given as

$$E = \frac{p^2}{2m} + \frac{1}{2}m\omega^2 x^2$$ (B.10)

Comparing (B.9) to (B.10), it is clear that the analogy is strong. If one takes L to be the "mass", then the charge plays the role of position and the flux plays the role of momentum.

Thus, it is clear that spatial confinement quantizes the electromagnetic wave into discrete modes. Then, each mode has a harmonic oscillator characterization in which the number of photons in that mode corresponds to the number of bosons in the particular harmonic oscillator. Generally, then, one can characterize the oscillator mode in terms of a photon Fock state (often called a number state) as

$$\psi_s \sim |n\rangle_s,$$ (B.11)

where n is the number of photons in that particular mode.

Returning now to QED and the photon-electron interaction, one can generally draw a so-called Feynman diagram as shown in Fig. B.2 for the emission of a photon by an electron. As described in the figure caption, the incident electron (coming from the left according to the arrow of time shown at the bottom) emits a photon and loses the corresponding energy and momentum. This interaction would be described by the corresponding set of operators as

$$a_{k_e-k_p}^\dagger b_{k_p}^\dagger a_{k_e},$$ (B.12)

which expresses the annihilation of the initial electron (the a term) and the creation of a photon (the central term) and an electron at the reduced energy and momentum. As usual, a operators are used for the electrons and b operators are used for the photon. This notation may be recognized in Chap. 2 where the Jaynes-Cummings model is described. Here, the order of operators is not particularly important, although the operations proceed

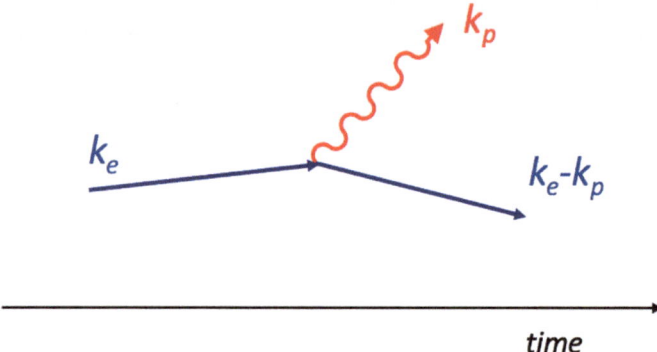

Fig. B.2 The so-called Feynman diagram for an electron emitting a photon. Here, an incident electron, with energy $E(k_e)$ and momentum k_e emits a photon of energy $\hbar\omega$ and momentum k_p. This results in the electron's final state (on the right) having an energy $E(k_e) - \hbar\omega$ and momentum $k_e - k_p$

from the right to the left. Although the figure would lead one to believe that this is an instantaneous interaction, this is only true on time scales of common usage. The actual interaction can take a few femtoseconds to occur, and this is much shorter than what is our accustomed scale. However, it is easily achieved by scientists exploring the details of such interactions, since laser pulses with time scales much shorter than this, particularly those generated by X-ray lasers [3] are used for studies of materials on this fast time scale.

References

1. Balanis, C. A.: Advanced Engineering Electromagnetics. (Wiley, New York, 2008).
2. Ferry, D. K.: Quantum Mechanics, 3rd Ed. (CRC Press, Boca Raton, 2021).
3. Maroju, P. K., Grazioli, C., Di Fraia, M., Moioli, M., et al.: Attosecond Pulse Shaping Using a Seeded Free Electron-Laser. Nature **578**, 386 (2020).